ESSENTIALS OF COORDINATION CHEMISTRY

ESSENTIALS OF COORDINATION CHEMISTRY
A Simplified Approach with 3D Visuals

VASISHTA BHATT
UGC-Human Resource Development Centre
Sardar Patel University
Vallabh Vidyanagar, Gujarat, India

Amsterdam • Boston • Heidelberg • London
New York • Oxford • Paris • San Diego
San Francisco • Singapore • Sydney • Tokyo
Academic Press is an imprint of Elsevier

Academic Press is an imprint of Elsevier
125 London Wall, London EC2Y 5AS, UK
525 B Street, Suite 1800, San Diego, CA 92101-4495, USA
225 Wyman Street, Waltham, MA 02451, USA
The Boulevard, Langford Lane, Kidlington, Oxford OX5 1GB, UK

Notices
Knowledge and best practice in this field are constantly changing. As new research and
experience broaden our understanding, changes in research methods, professional
practices, or medical treatment may become necessary.

Practitioners and researchers must always rely on their own experience and knowledge in
evaluating and using any information, methods, compounds, or experiments described
herein. In using such information or methods they should be mindful of their own safety
and the safety of others, including parties for whom they have a professional
responsibility.

To the fullest extent of the law, neither the Publisher nor the authors, contributors, or
editors, assume any liability for any injury and/or damage to persons or property as a
matter of products liability, negligence or otherwise, or from any use or operation of any
methods, products, instructions, or ideas contained in the material herein.

ISBN: 978-0-12-803895-6

Library of Congress Cataloging-in-Publication Data
A catalogue record for this book is available from the Library of Congress

British Library Cataloguing-in-Publication Data
A catalogue record for this book is available from the British Library

For information on all Academic Press publications
visit our website at http://store.elsevier.com/

Working together
to grow libraries in
developing countries

www.elsevier.com • www.bookaid.org

To the teachers and learners of coordination chemistry

CONTENTS

PREFACE

This book is an outcome of my research programmes, coupled with teaching about 10,000 learners over a period of two decades. While teaching, I realized that students perceive chemistry as one of the most difficult subjects. On pondering over the reason, I came to a conclusion that teachers find it difficult to make the students imagine the structures of atoms, ions and molecules as they are to be understood. As a result, many chemistry classrooms lack the desired resonance. The obvious remedy was to create the images and movies that describe these species to a maximum extent. I have been trying to achieve this end for some time now. I realized that there is still more to be done to make chemistry simple enough for students.

Apart from this, there are other constraints faced by chemistry teachers as well, which require attention. Often, professors face time constraints in discussing the intricacies of certain concepts in the classroom. Hence, it becomes a 'fixed place—fixed time' mode of education. In this age, when knowledge could transcend the borders of time, place and community, an all times—all places mode of learning was, therefore, an inevitable way out. This book is an outcome of the above rationale.

As it is said, a book is a person's best friend. Even more so, if it offers self-learning facilities with features like numerous images and videos of good quality, clear illustrations and relevant citations, as well as lucid language.

Essentials of Coordination Chemistry: Simplified Approach with 3D Visuals is a step in the direction of developing a lifelong friendship between the book and the learner. It includes subject matter for an entire semester in the course of BS/BSc with chemistry as a major. It begins with basic coordination chemistry and basic concepts of symmetry, which serve as a primer for advanced learning. A detailed discussion on isomerism amongst coordination compounds is provided in the subsequent chapter. The next three chapters cover classical topics related to the thermodynamics and kinetics of complex formation, as well as a variety of reactions occurring in coordination compounds of square planar and octahedral geometry. In the last section of the book, beginning with basic organometallic chemistry, a brief discussion on general organometallic chemistry is provided, which is then extended particularly to an industrially important class of organometallic complexes: metal carbonyls and metal nitrosyls. The electronic version of this is expected to be a camera-ready set of lectures that can be readily delivered in the class. Moreover, at the end of every chapter, some exercises based on the text are also provided. An attempt has been made to include the latest developments in the subject. For instance, brief information on the Hoveyda—Grubbs' catalyst or the Suzuki reaction, as well as others from the twenty-first century, is included in this book. I hope that this book will cater to the needs of the students and teachers.

I am indeed keen to receive constructive comments and suggestions from the teachers' and learners' fraternity for potential improvisations in this book.

Vasishta Bhatt

ACKNOWLEDGEMENTS

First of all, I must thank Elsevier Inc. for putting trust in my ability to complete this work. I thank Cathleen Sether (Publishing Director), Katey Birtcher (Senior Acquisitions Editor, Chemistry), Jill Cetel (Senior Editorial Project Manager) and the entire Elsevier team for making this dream journey a reality.

I sincerely acknowledge the support of the University Grants Commission (UGC), New Delhi, India, for giving me an office and a position from which I could comfortably complete this book; it would have been impossible otherwise. The Vice Chancellor of Sardar Patel University, Dr Harish Padh, has remained a prime source of my inspiration, as he is a role-model scientist anyone would want to emulate. Dr A.R. Jani, the Director of UGC-Human Resource Development Centre at Sardar Patel University, was very happy when I took this assignment, and he readily arranged for all the facilities that would make me comfortable in preparing this book. He also provided vital scientific guidance. My fellow colleague, Mr Vimal Shah, needs to be acknowledged for his valuable suggestions; as a layman, it really helped me in predicting the learner's perspective on technical issues. The entire staff of my institute whole heartedly supported me in this endeavour. Prof. Mangala Sunder Krishnan, IITM Chennai and National Coordinator of the NPTEL, has remained the torch bearer in this entire journey. The faculty members of the chemistry department of Sardar Patel University, namely Prof. M.N. Patel, Prof. N.V. Sastry and Prof. D.K. Raval, strongly encouraged me and offered all the guidance that I desired. How can I forget the assistance provided by Mr Ravish Patel, who worked as my assistant and helped me on each and every step!

My family remained the main driving force behind this project. I could never have completed this book if they were not there. Sonal (my wife) and Ishan (my son) contributed in all possible manners, beginning with the sacrifice of their vacation outings. Both of them gave creative criticism and suggestions on many pedagogical issues whenever I was in doubt. Mothers and fathers are the first teachers of a child; Ramaben and Dineshbhai will be the happiest people on the earth when this book takes its final form.

All the teachers and learners of coordination chemistry have knowingly or unknowingly contributed a lot to this work. This book is now to be tested for their expectations.

Vasishta Bhatt

ACKNOWLEDGEMENTS

CHAPTER 1

Basic Coordination Chemistry

Contents

1. INTRODUCTION

The coordination compounds found their applications long before the establishment of coordination chemistry. Bright red coloured alizarin dyes were under applications even before the fifteenth century. This bright red dye, now characterized as a chelated complex of hydroxyanthraquinone with calcium and aluminium metal ions, is shown in Figure 1.

Later, in the sixteenth century, the formation of a well-known member of today's coordination chemistry family, the tetraamminecupric ion

Essentials of Coordination Chemistry
http://dx.doi.org/10.1016/B978-0-12-803895-6.00001-X

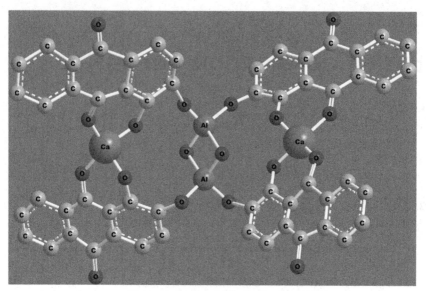

Figure 1 Structure of alizarin dye.

$[Cu(NH_3)_4]^{+2}$, was recorded upon contact between brass alloy and ammonium chloride. Addition of Prussian blue $Fe_4[Fe(CN)_6]_3 \cdot xH_2O$ increased the use of coordination compounds in dyes and pigments. A platinum complex $K_2[PtCl_6]$ offered an application for the refinement of platinum metal. Thus, before the coordination chemistry was structured, the coordination compounds, complexes and chelates found their applications.

A systematic investigation of structure and bonding in coordination chemistry began with the inquisitiveness of Tassaert (1798), which was extended by distinguished chemists like Wilhelm Blomstrand, Jorgensen and Alfred Werner [1] until the end of the nineteenth century. In the events, Werner's coordination theory (1893) became the base of the modern coordination chemistry. It is worth noting that the electron was discovered subsequent to Werner's theory.

The bonding in compounds like $CoCl_3$ and NH_3 were easily understood and explained and hence such compounds were regarded as simple compounds. For instance, the $+3$ formal oxidation of cobalt in cobalt chloride is balanced by three uni-negative chloride ions and the coexistence of these ionic moieties to form a molecule is understood and explained. Similarly, the valence shell ($n = 2$) of nitrogen ($N = 7$) contains five electrons and four orbitals (2s, $2p_x$, $2p_y$ and $2p_z$). Keeping an electron pair in one of these orbitals while the other three remains half filled, an opportunity for three hydrogen atoms to contribute one electron each for the formation of a covalent bond with nitrogen, can also be explained. Thus an ammonia molecule has three N—H covalent bonds and one lone pair of electrons over the nitrogen atom. It is worth noticing here that all the valencies of all the atoms in both the molecules are fully satisfied and hence there is no further scope of bonding.

A 'complex' situation arises when one comes to know that the molecule $CoCl_3$ can encompass six ammonia molecules, resulting into a third independent entity. This situation was fully understood and explained by Werner's coordination theory, followed by naming the entity as 'complex'.

1.1 Definitions

Coordination compounds are the compounds containing one or more coordinate covalent bonds.

Coordinate covalent bonds are the covalent bonds in which both the bonding electrons are contributed by one of the bond partners. Figure 2 distinguishes the covalent bonds from the coordinate covalent bond in NH_3BF_3. While the three B—F covalent bonds are formed due to the sharing of electron pairs resulting from contributions of both boron and fluorine atoms, an N—B bond is formed due to the donation of a lone pair

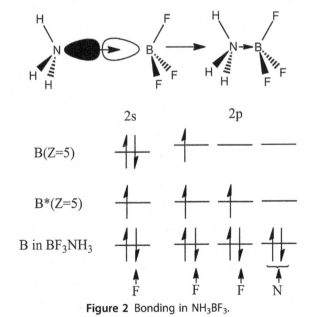

Figure 2 Bonding in NH_3BF_3.

of electrons from nitrogen into the empty orbitals of boron. The coordinate covalent bond is shown by an arrow with its head pointing towards the direction of the donation of an electron pair, as shown in Figure 2.

A **complex** is a molecule/ion containing a central metal atom/ion surrounded by a definite number of ligands held by secondary valences or coordinate covalent bonds.

Primary valency refers to the charge over the metal ion e.g. Co(III) has +3 charge, which can be balanced by −3 charge-forming compounds like $CoCl_3$. The primary valency is ionic and is satisfied in the second coordination sphere, as shown in Figure 3.

Secondary valency is the number of empty valence orbitals, as illustrated for $[Co(NH_3)_6]Cl_3$ in the figure. The Co(III) ion has six empty valence orbitals. Hence its secondary valency is six. Secondary valency is a coordinate covalent valency, and it is satisfied in the first coordination sphere of the metal ion, as shown in Figure 4.

Coordination number is a property of the metal ion representing the total number of donor atoms directly attached to the central atom. In the above case, the coordination number of Co(III) is six, as six nitrogen donor atoms are directly connected to the central metal ion (cobalt(III)).

Ligand is any atom, ion or neutral molecule capable of donating an electron pair and bonded to the central metal ion or atom through secondary valency.

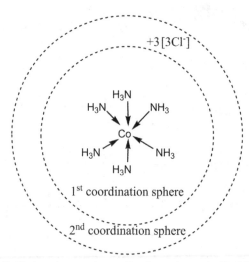

Figure 3 First and second coordination spheres in $[Co(NH_3)_6]Cl_3$.

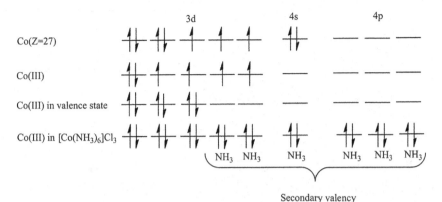

Secondary valency

Figure 4 Secondary valency of Co(III) in [Co(NH$_3$)$_6$]Cl$_3$.

Dentate character is a property of a ligand representing a number of coordinating atoms.

In the case of [Co(NH$_3$)$_6$]Cl$_3$, ammonia, NH$_3$ the ligand contains one donor atom (N). Hence its dentate character is one and is classified as a monodentate ligand. Similarly, chloro (Cl$^-$) is an anionic, monoatomic and monodentate ligand, while hydroxo (OH$^-$) is a diatomic, monodentate and anionic ligand. Aquo (OH$_2$) represents a neutral triatomic monodentate ligand. A few popular ligands and their characteristics are shown in Figure 5.

Due to a higher dentate character of ligands, a variety of complexes known as chelate is also formed sometimes. **Chelate** is a compound formed when a polydentate ligand uses more than one of its coordinating atoms to form a closed-ring structure, which includes the central metal ion. Five- and six-membered rings are known to provide extra stability to the chelates. The process of chelate formation is known as chelation. A polydentate ligand involved in chelate formation is also known as a chelating ligand. Chelates generally exhibit higher stability than analogous complexes.

A polydentate ligand may be attached to the central metal ion through more than one kind of functional group. The number and kind of linkages by which the metal ion is attached with the ligands can thus become a criterion for the classification of chelates. The covalent bonds are formed by the replacement of one or more H–atoms, while coordinate covalent bonds are formed by the donation of an electron pair from the ligands. Some of the chelates involving a variety of polydentate ligands and linkages are shown in Figure 6. The coordinate covalent linkages are shown by thin, thread–like bonds.

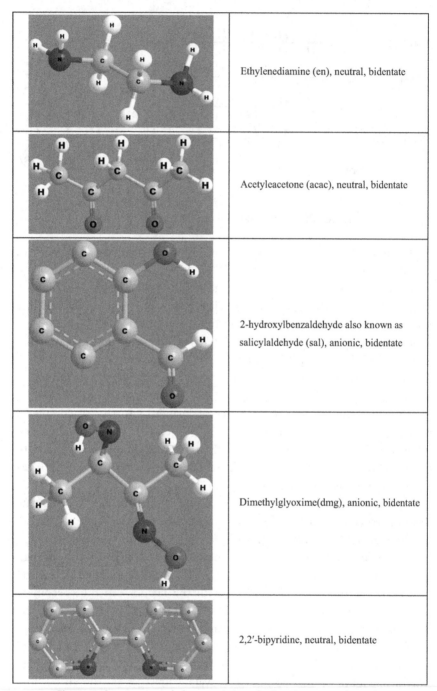

	Ethylenediamine (en), neutral, bidentate
	Acetyleacetone (acac), neutral, bidentate
	2-hydroxylbenzaldehyde also known as salicylaldehyde (sal), anionic, bidentate
	Dimethylglyoxime(dmg), anionic, bidentate
	2,2'-bipyridine, neutral, bidentate

Figure 5 Structures and characteristics of a few important ligands.

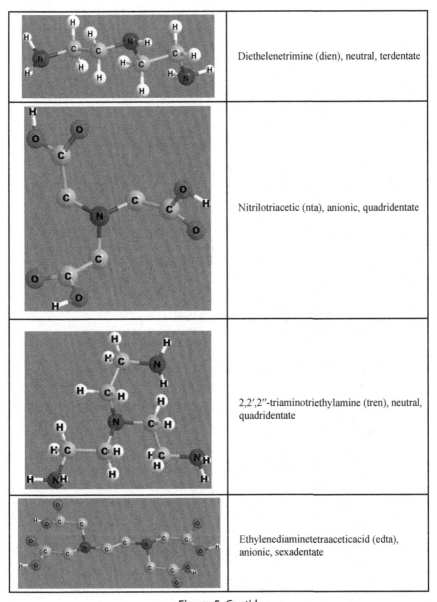

	Diethelenetrimine (dien), neutral, terdentate
	Nitrilotriacetic (nta), anionic, quadridentate
	2,2',2''-triaminotriethylamine (tren), neutral, quadridentate
	Ethylenediaminetetraaceticacid (edta), anionic, sexadentate

Figure 5 Cont'd

EXAMPLE	CHARACTERISTICS
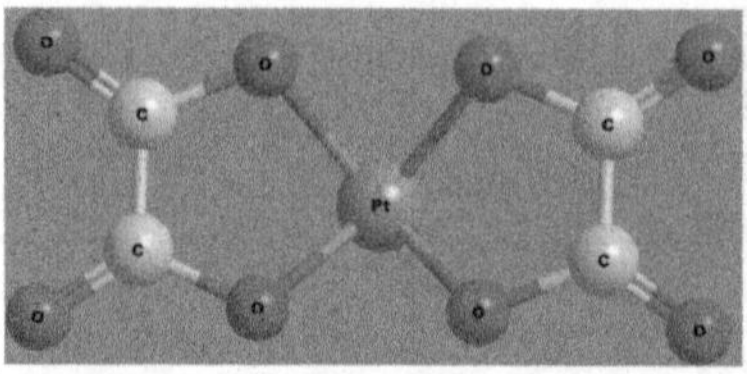	Oxalic acid Anionic Bidentate Two covalent bonds Two five-membered rings in the chelate
	Glycine Anionic Bidentate One covalent and one coordinate covalent bond Two five-membered rings in the chelate
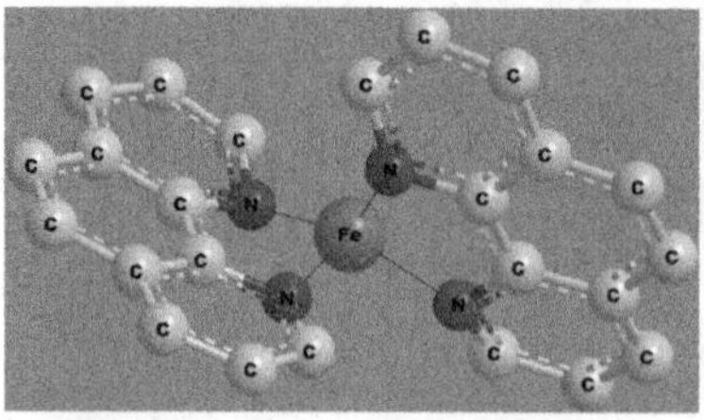	1,10-phenanthroline Neutral Bidentate Two coordinate covalent bonds Two five-membered rings in the chelate
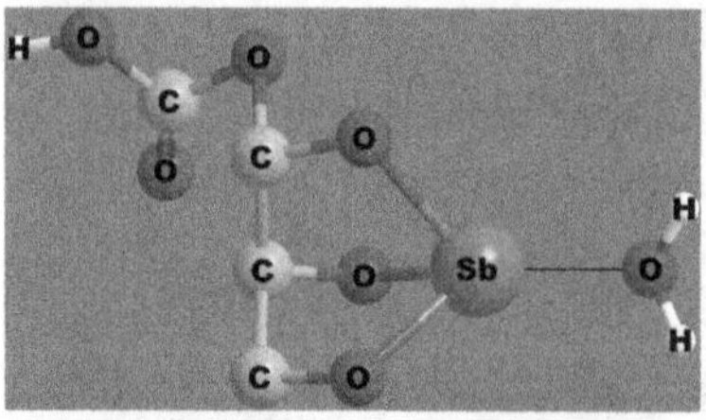	1,2,3-trihydroxypropyl hydrogen carbonate Anionic Tridentate Three covalent bonds Two five-membered rings and one six-membered ring in the chelate
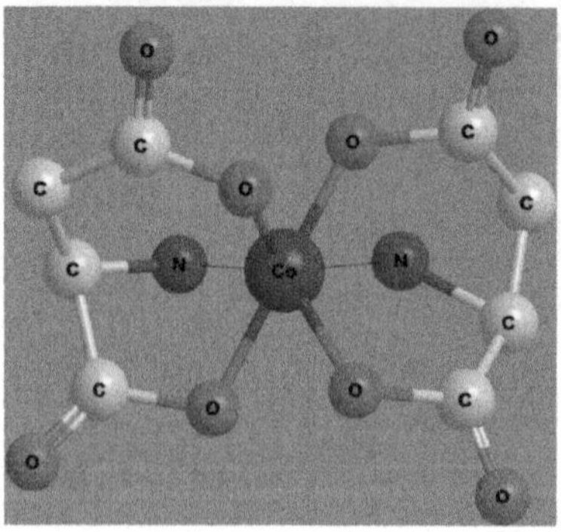	2-aminosuccinic acid also known as aspartic acid Anionic Tridentate Two covalent and one coordinate covalent bond Two five-membered rings, two six-membered rings and two seven memebered rings in the chelate

Figure 6 Structures and characteristics of a few chelates.

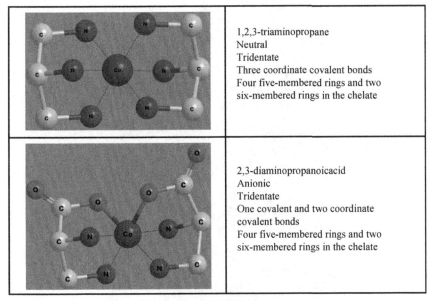

	1,2,3-triaminopropane Neutral Tridentate Three coordinate covalent bonds Four five-membered rings and two six-membered rings in the chelate
	2,3-diaminopropanoicacid Anionic Tridentate One covalent and two coordinate covalent bonds Four five-membered rings and two six-membered rings in the chelate

Figure 6 Cont'd

Polynuclear complex is a complex with more than one metal atom/ion. These metal ions are sometimes bridged through appropriate ligands, resulting into the formation of a bridged polynuclear complex.

2. NOMENCLATURE

A systematic nomenclature of coordination compounds requires a careful consideration of the following rules [2]. The learner should first learn all of these rules by heart and then do sufficient practice to master the procedure.

2.1 For Writing the Coordination Formula

1. Place the symbol of the central atom first followed by the symbol of the ligand in the following order: anionic, neutral and cationic. Enclose the complex in a square bracket.
2. If the formula of a charged complex is written without any counter-ion, the charge is to be indicated outside the square bracket as a right superscript with the number preceding the sign, as in $[PtCl_6]^{2-}$ or $[Cr(OH_2)_6]^{3+}$; the oxidation number of a central atom/ion may be optionally represented by a Roman numeral placed as a right superscript on the element symbol, as in $[Cr^{III}Cl_3(OH_2)_3]$ and $[Fe^{II}(CO)_4]^{2-}$.

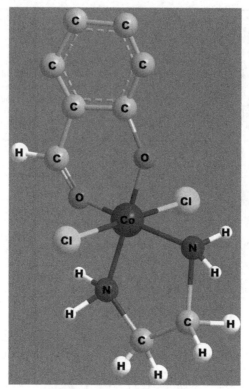

Figure 7 *trans*-[Co^{III}Cl$_2$(sal)(en)].

3. The anionic ligands are cited first and in alphabetical order, according to the first symbols of their formulae.

4. The neutral and cationic ligands are then listed in the following order: H_2O, NH_3, other inorganic ligands and organic ligands in alphabetical order.

5. The structural information may be given by prefixes such as *cis-*, *trans-*, *fac-*, *mer-* etc.

To illustrate the above rules, consider the writing of a coordination formula for the compound shown in Figure 7 as *trans*-[Co^{III}Cl$_2$(sal)(en)].

N.B.: The ligand abbreviations and the formulae of the polyatomic ligands are generally placed in parentheses.

2.2 For Writing the Names

1. The cation is named first, followed by the name of the anion irrespective of whether the cation or the anion is the complex species.

2. In a complex (cationic, anionic or neutral), the names of the ligand are cited alphabetically irrespective of their charge without separation, and the name of the central metal atom is the last.
3. The names of the anionic complexes have the specific ending '-ate' or '-ic' (if named as acid), while there is no such specific ending of cationic or neutral complexes.
4. The oxidation state of the central atom is indicated by a Roman numeral in the parenthesis at the end of the name of the complex.
5. Name of the ligands: the specific ending '-o' is given to organic/ inorganic anionic ligands. In the case of ligand names ending in '-ide', '-ite' or '-ate', 'e' is replaced by 'o', giving '-ido', '-ito' or '-ato' respectively, as in -azide (N_3^-) becomes azido, -nitrite (NO_3^-) becomes nitrito and -sulphate (SO_4^{-2}) becomes sulfato.

   However, certain anionic ligands are exceptions to the above rule and are named as shown in Table 1

   There is no change in the name of neutral ligands. They are named as a molecule as in $NH_2CH_2CH_2NH_2$ (en) as ethylenediamine and $NH_2CH_2CH_2NH\ CH_2CH_2NH_2$ (dien) as diethylenetriamine. Here also, the exceptions are H_2O (named as aquo), NH_3 (as ammine), NO (as nitrosyl) and CO (as carbonyl) etc.

   The cationic ligands have the specific ending 'ium', as in $NH_2-NH_3^+$, named as hydrazinium, and $H_2N-CH_2-CH_2-NH_3^+$, named as 2-aminoethylammonium.
6. The number of each kind of simple ligand is indicated by prefixes such as mono-for one ligand, which is usually omitted, di- for two ligands and tri- for three ligands without any space. In the case of complicated ligands, the prefixes bis- for two ligands, tris- for three ligands and tetrakis- for four ligands (with the name of the ligand enclosed in parenthesis) is used.
7. The coordinating atom of a ligand to the central atom is indicated by placing the symbol κ (Kappa) followed by the elemental symbol after

Table 1 Names used for anionic ligands according to International Union of Pure and Applied Chemistry (IUPAC) nomenclature

Ligand	Name	Ligand	Name	Ligand	Name
F^-	Fluoro	OH^-	Hydroxo	O_2^{2-}	Peroxo
Cl^-	Chloro	CN^-	Cyano	CH_3O^-	Methoxo
Br^-	Bromo	HS^-	Thiolo	$C_5H_5^-$	Cyclopentadienyl
I^-	Iodo	S^{2-}	Thio	$C_6H_5^-$	Phenyl
H^-	Hydrido	O_2^-	Superoxo	O^{2-}	Oxo

the name of the ligand, as in M-SCN-, it is written as thiocyanato-κS and in M-NCS-, it is written as iosthiocynato-κN.

8. A ligand that bridges two central atoms is designated by the symbol μ before its name.

Here are a few examples of using IUPAC nomenclature for writing the formulae and names of coordination complexes.

Formula	Name
$[Fe(CN)_6]^{4-}$	Hexacyanoferrate(II) ion
$[CoCl_3(NH_3)_4]$	Triamminetrichlorocobalt(III)
$[CuCl_2(CH_3NH_2)_2]$	Dichlorobis(methylamine)copper(II)
Cis-$[PtCl_2(NH_3)_2]$	Cis-diamminedichloroplatinum(II)
$[PtCl(en)NH_2NH_3]^{+2}$	Chloroethylenediaminehydraziniumplatinum(II) ion
$[Fe(CN)(CNCH_2C_6H_5)_5]^+$	Pentakis(benzylisocynide)cyano iron(III) ion
$[Al(OH)(H_2O)_5]^{+2}$	Pentaaquohydroxoaluminum(III) ion
$[Co(CN)(CO)_2NO]^-$	Dicarbonylcyanonitrosylcobaltate(0) ion
$[PtClNO_3(NH_3)_2(en)]SO_4$	Diammine(ethylenediamine) chloronitroplatinum(IV) sulphate
$K[SbCl_5(C_6H_5)]$	Potassium pentachloro(phenyl) antimonite(V)
$H_2[PtCl_6]$	Hexachloroplatinic(IV) acid
$Na_2[Fe(CN)_5NO]$	Sodium pentacyanonitrosylferrate(III)
$[PtPy_4][PtCl_4]$	Tetrakis(pyridine)platinum(II) tetrachloroplatinate(II)
$[Pt(NH_3)_4Br_2]Br_2$	Tetramminedibromoplatinum(IV) bromide

3. THEORIES OF BONDING IN COORDINATION COMPOUNDS

The theories of bonding in coordination compounds [3] have evolved subsequent to Werner's coordination theory (1893). Werner introduced the concept of primary and secondary valency, explaining the formation of the coordination compounds. The 18-electron rule, stating that the stable complexes with low formal oxidation states of metal ions should have 18 bonding electrons around the metal ion, became an important beginning point toward the study of the stability of the complexes. The 18-electron rule is significant in modern coordination chemistry as it is also supported by the molecular orbital theory. However, a smaller number of complexes with metals in low oxidation states restrict its wide applicability. An important advance in the theories of bonding in coordination compounds was the introduction of

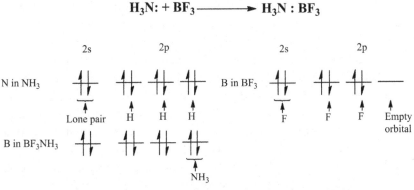

Figure 8 Covalent and coordinate covalent bonding in BF_3NH_3.

valence bond theory (VBT), which is actually an extension of Lewis theory (1902) and Heitler–London theory (1927). From the above two theories, the sharing of electron pairs by atoms and the formation of bonds was understood and explained. The VBT developed by Linus Pauling introduced two key concepts in the theories of bonding, viz, the hybridization of orbitals and resonance, the absence of which were the drawbacks of the previous theories.

According to this theory, the metal ions are regarded as Lewis acids characterized by the availability of low lying, empty orbitals suitable for accommodating the accepted electrons. The ligands are the Lewis bases, characterized by the availability of a lone pair of electrons that can be readily donated, resulting into a formation of a coordinate covalent bond. Thus, the acid–base characteristics of metal and ligand complement each other to give rise to a coordinate covalent bond, as shown in Figure 8.

Hybridization of atomic orbitals, a mathematical tool for mixing the atomic orbitals to give an equal number of hybrid orbitals with superior directional properties, became a very important feature of this theory in explaining various geometries observed in coordination compounds.

4. GEOMETRIES OF COMPLEXES WITH DIFFERENT COORDINATION NUMBERS

1. **Coordination number 2:** This is a relatively uncommon coordination number and is restricted to a few cases, such as Cu(I), Ag(I), Au(I), and Hg(II) ions having d^{10} configuration. $[CuCl_2]^-$ and $[Ag(NH_3)_2]^+$ are the representative complexes of this coordination number.

The possible geometries for the complexes with CN = 2 are linear and angular. Linear geometry is commonly observed in such complexes, as it involves minimum ligand–ligand repulsion. These complexes involve sd-hybridization between the central metal atom orbitals, as shown in Figure 9.

Orgel has suggested that $(n-1)$d orbitals have nearly the same energy as that of ns and np orbitals. The d_{z^2} orbitals can enter into this hybridization to remove electron density away from the region of ligands and stabilize the complex. Initially '4s' and '3d$_{z^2}$' orbitals hybridize to give two sd-hybrid orbitals, namely ψ_1 and ψ_2. The ψ_2 sd-orbitals have their positive lobes along the z axis, while the ψ_1 is in the XY plane. The ψ_2 orbitals, further hybridized with p_z orbitals, give the hybrid orbitals concentrated along the z-direction. The electron pair from the d_{z^2} orbitals now occupy the sd orbital XY plane, while hybrid orbitals concentrated along the z-direction are available for a stronger bond along the z axis, resulting into a complex with linear geometry.

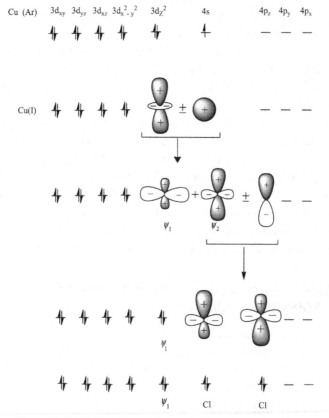

Figure 9 Hybridization explaining the bonding in $[CuCl_2]^-$.

The chelates of this coordination number are less stable than analogous complexes. Consider the complexes $[Ag(en)]^+$ and $[(Ag(NH_3)_2]^+$. In these complexes, the favoured geometry is linear. In $[Ag(en)]^+$ chelate, a five-membered ring is formed, and the 'en' molecule cannot occupy linear coordination positions around Ag^+ without introducing strain in the chelate ring, which destabilizes the complex. Hence $[Ag(en)]^+$ is less stable than $[(Ag(NH_3)_2]^+$. The Ag(I) chelates with six-, seven- or eight-membered rings are more stable; as in larger-sized chelate rings, the strain is less, hence stability is greater.

2. **Coordination number 3:** This is very rare coordination number. In most cases, detailed studies have shown a higher coordination number attained by the central metal atom through dimerization and bridging, as in the cases of $AlCl_3$ and $PtCl_2(PR_3)$, shown in Figure 10.

    However, CN = 3 is also illustrated by a few true complexes such as HgI_3 and $[Cu(SPMe_3)]$ ions. These complexes exhibit trigonal planar geometry, as shown in the figure.

3. **Coordination number 4:** This is one of the most commonly observed coordination numbers. There are two possible geometries associated with CN = 4, viz, tetrahedral and square planar.

    a. **Tetrahedral complexes:** Tetrahedral geometry is fairly common in four-coordinate complexes of nontransition metal ions. This geometry involves sp^3 hybridization of orbitals of the central atom. This geometry is favoured by large ligands like Cl^-, Br^-, I^- and small metal ions like Be^{+2} and Al^{+3}. Complexes like $[BeF_4]^{2-}$, $[ZnCl_4]^{2-}$ and $Be(acac)_2$, shown in Figure 11, illustrate the tetrahedral complexes with CN = 4.

    b. **Square planar complexes:** Square planar coordination is less favoured structurally (due to stronger ligand–ligand repulsion) than tetrahedral coordination, especially by large ligands. If the ligands are small, octahedral coordination can be achieved. Few metal ions are known to form square planar complexes; it is mostly restricted to d^8 to d^9 ions like Pt(II), Pd(II), Ni(II), Cu(II) and Au(II), as illustrated in $[Pt(NH_3)_4]^{2+}$, $[PtCl_2(NH_3)_2]$ and $[Ni(CN)_4]^{2-}$ shown in Figure 12.

4. **Coordination number 5:** The two possible geometries for this coordination number are trigonal bipyramidal (TBP) and square pyramidal (SqPy). In most cases, the structure is found to be distorted. The conversion between these two geometries can easily occur by slight

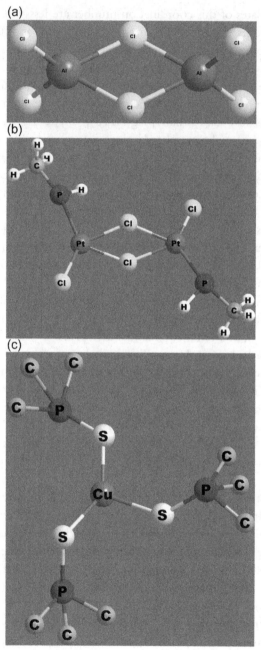

Figure 10 Dimerization and bridging in (a) AlCl$_3$ (b) PtCl$_2$P(CH$_3$) and (c) [Cu(SPMe$_3$)].

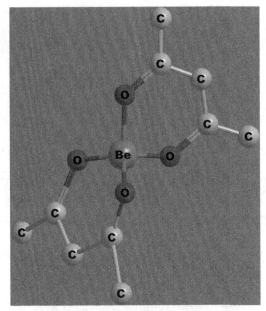

Figure 11 Tetrahedral Be(acac)$_2$ chelate.

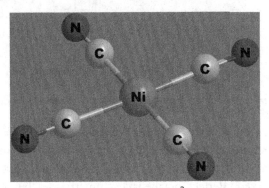

Figure 12 Square planar [Ni(CN)$_4$]$^{2-}$ complex ion.

deformation. However, [CuCl$_5$]$^{-3}$ and [NiCN$_5$]$^{-3}$ exhibit TBP and square pyramidal geometry respectively, as shown in Figure 13.

5. **Coordination number 6:** This is the most common coordination number. Some metals like Cr(III) and Co(III) almost exclusively form six-coordinate complexes. There are three popular arrangements of six ligands around the central metal ion, viz, hexagonal planar, trigonal prismatic and octahedral, as shown in Figure 14.

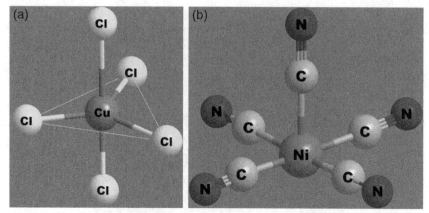

Figure 13 (a) Trigonal bipyramidal [CuCl$_5$]$^{-3}$ and (b) Square pyramidal [NiCN$_5$]$^{-3}$ ions.

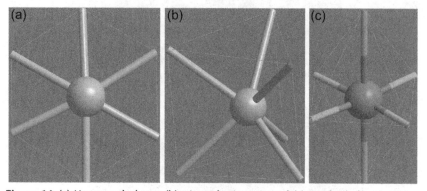

Figure 14 (a) Hexagonal planar, (b) trigonal prismatic and (c) octahedral geometries.

An extensive structural study of the complexes with CN = 6 has shown that the arrangement of six ligands in a six-coordination compound is always octahedral. In octahedral complexes, d^{2}sp^3 or sp^{3}d^2 hybridization of orbitals is observed. Perfect octahedral symmetry can be observed only when all of the six ligands are identical, otherwise the geometry gets distorted. It may show either tetragonal or trigonal distortion, as shown in Figure 15.

Complexes with higher coordination numbers have also been reported, but considering the scope of the book, the discussion on such complexes is knowingly omitted.

Though VBT has remained a necessary tool for the study of metal complexes, it is not sufficient to explain the colour and characteristics of the absorption spectra of the complexes. The theory could not demystify the

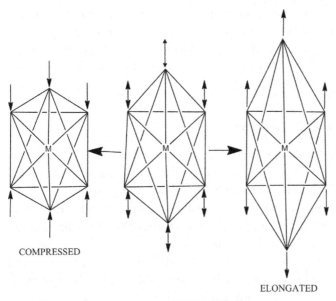

COMPRESSED

ELONGATED

**TRIGONAL
DISTORTION**

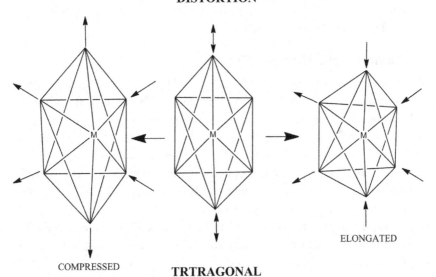

COMPRESSED

ELONGATED

**TRTRAGONAL
DISTORTION**

Figure 15 Distortions in octahedra.

mystery of the formation of outer and inner orbital complexes. The distortions in the shapes of the coordination complexes from regular geometry were outside the purview of the VBT. Moreover, orbital contributions and temperature dependency on magnetic moment of coordination complex are not properly explained by VBT.

5. CRYSTAL FIELD THEORY

Bethe and Van Vleck developed the crystal field theory (CFT) [4] for coordination in the early 1930s. CFT treats the coordination compounds as simple ionic compounds. The bonding in coordination compounds is due to electrostatic forces between the positively charged metal ions and the negatively charged ligands and, in the case of neutral ligands, dipoles. This theory fits well for complexes of metal ions with small and highly electronegative ligands, such as F^-, Cl^- and OH_2, but it does not work well with ligands having less polarity, such as carbon monoxide. CFT also needs modification to explain the difference between the spectra of free metal ions from the complexed ones. Considering the simplicity of the CFT and the complexity of the bonding in complexes, the CFT is regarded as a theory that is too good to be true.

5.1 Ligand Field Theory (LFT)

The first article on Ligand Field Theory (LFT) [5] was authored by Orgel and Griffith in 1957. The LFT considers the contributions from both ionic and covalent bonding for accounting the properties of coordination compounds. The LFT is a combination of molecular orbital (MO) theory and CFT. The transition metals form a large number of complexes: they formed coloured ions, they form paramagnetic compounds, they show variable oxidation states and they also exhibit catalytic activity. All these properties can only be explained by studying their electronic structures, spectra and magnetic properties. LFT is a term that is used to describe the whole toolkit required to understand the bonding and other properties of the transition metal elements and their compounds.

6. SPLITTING OF d-ORBITALS IN FIELDS OF DIFFERENT GEOMETRY

The fivefold degenerate d-orbitals are shown in Figure 16. They are designated as d_{xy}, d_{yz}, d_{xz}, $d_{x^2-y^2}$ and d_{z^2}. The lobes of d_{xy}, d_{yz} and d_{xz}

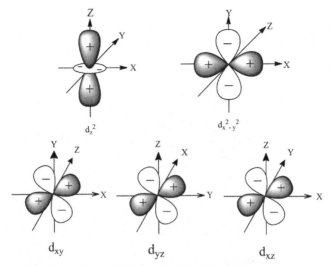

Figure 16 Shapes of five d-orbitals.

orbitals lie in the xy, yz and xz planes, respectively. These lobes do not touch any of the axes. These orbitals are known as a t_{2g} set of orbitals. The lobes of $d_{x^2-y^2}$ and d_{z^2} lie on the x and y axes and z axis, respectively. These orbitals are known as an e_g group of orbitals.

Figure 17 shows the nucleus of a transition metal ion coinciding the origin (0,0,0) of a Cartesian coordinate system. The five d-orbitals of the ion are thus extending from the origin of the Cartesian coordinate system. Consider six uni-negative monodentate ligands slowly approaching the central metal ion, as indicated by negative point charges. The ligands are approaching along the x, y and z axes, forming an octahedral field. As the ligands come closer to the nucleus of the metal ion, they encounter the orbitals of the metal ion. Thus there will be repulsion between the orbital (they contain electrons) and the ligands (they are negatively charged). The e_g orbitals that are oriented along the axes face the stronger repulsion, as compared to that of the t_{2g} orbitals lying in the plane. Due to this, the degeneracy of the fivefold d-orbitals split into t_{2g} orbitals with lower energy and e_g orbitals with higher energy, as shown in the left part of Figure 18.

The magnitude of the splitting of d-orbitals due to the introduction of an octahedral ligand field is Δo, where $3/5\ \Delta o$ is the extent of increase in energy of e_g orbitals and $2/5\ \Delta o$ is the corresponding decrease in energy of t_{2g} orbitals, as compared to the energy for the free metal ion.

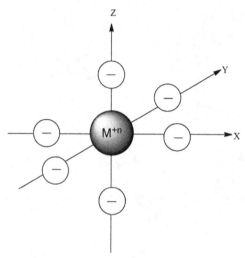

Figure 17 A metal ion coincident with the origin of a Cartesian coordinate system.

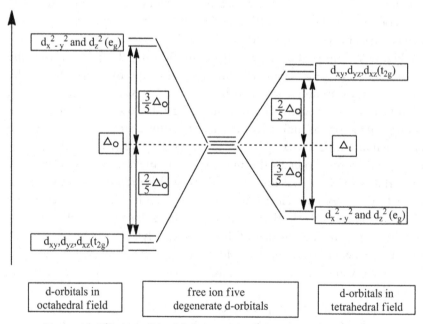

Figure 18 Splitting of d-orbitals in octahedral and tetrahedral fields.

Introduction of a tetrahedral ligand field can be visualized by coinciding the centre of a cube with the origin of the Cartesian coordinate system. The four monodentate ligands in a tetrahedral complex approach the nucleus of the metal ion through the alternate diagonals of the opposite faces of the cube. Thus these ligands approach the metal ion along the planes xy, yz and xz. This results into an increase in repulsion in t_{2g} orbitals and, in turn, the increase in energy of t_{2g} orbitals and corresponding decrease in the energy of e_g orbitals, as shown in the right part of Figure 18.

The magnitude of the splitting of d–orbitals due to introduction of a tetrahedral ligand field is Δt, where $2/5$ Δt is the increase in energy of t_{2g} orbitals, while $3/5$ Δt is the corresponding decrease in energy of e_g orbitals, as compared to the energy for the free metal ion.

If the ions and interionic distances are the same, an empirical relation between Δo and Δt is $\Delta t = 4/9 \, \Delta o$.

As Figure 18 shows the splitting of d–orbitals, it is evident that for a d^{10} system in an octahedral or a tetrahedral field, the net energy change due to splitting is zero.

Consider a d^{10} system in an octahedral field where the configuration becomes $t_{2g}^6 e_g^4$. Now the decrease in energy due to six electrons occupying t_{2g} is given by $6 \times (2/5)\Delta o = (12/5)\Delta o$, whereas the increase in energy due to four electrons occupying e_g orbitals is $4 \times (3/5)\Delta o = (12/5)\Delta o$. Thus for a spherically symmetrical system, the algebraic sum of all the energy shift of all orbitals is zero. The splitting pattern, in which the algebraic sum of the entire energy shift is zero, is said to obey the centre of gravity rule. Thus, the 'centre of gravity' is preserved in the splitting of d–orbitals.

6.1 Ligand Field Stabilization Energy (LFSE)

The splitting of d–orbitals results into the lowering of energy of some of the orbitals as compared to their equilibrium positions (t_{2g} orbitals in an octahedral field). Each of the electrons occupying such orbitals results into the lowering of energy. When the number of electrons occupying the low energy orbitals is greater than that of the higher energy orbitals, there is a net lowering of the energy of the system. This energy is called ligand field stabilization energy (LFSE).

The expression for LFSE can be obtained as follows:

Let us consider an ion in an octahedral ligand field. Hence, the d–orbitals will split into t_{2g} with lower energy ($2/5\,\Delta o$) and e_g with higher energy $(3/5\ \Delta o)$. Moreover, the amount of energy required for spin pairing of the electrons will be 'P'.

Thus, LFSE in an octahedral ligand field can be given as the following:

$$\text{LFSE} = \text{Number of electrons in } t_{2g} \times \left(-\frac{2}{5}\Delta o\right) + P$$

$$+ \text{Number of electrons in } e_g \times \left(\frac{3}{5}\Delta o\right) + P$$

While LFSE in a tetrahedral ligand field can be given as the following:

$$\text{LFSE} = \text{Number of electrons in } e_g \times \left(-\frac{3}{5}\Delta t\right) + P$$

$$+ \text{Number of electrons in } t_{2g} \times \left(\frac{2}{5}\Delta t\right) + P$$

6.1.1 Tetragonal Distortions in an Octahedral Field

Distortions in geometries can further split the d-orbitals. Consider the two common distortions in an octahedral field:

1. Slowly removing two trans-ligands lying on the z axis.
2. Compressing two trans-ligands lying on the z axis.

Slowly removing two trans-ligands lying on the z axis:

Upon removal of two trans-ligands lying on the z axis, the repulsion faced by the d_{z^2} orbital decreases significantly, and also a slight reduction in repulsion is experienced by t_{2g} orbitals with the z-component (i.e. d_{xz} and d_{yz}), as shown in Figure 19. Thus as expected, the e_g orbitals further split into d_{z^2} with lower energy and $d_{x^2-y^2}$ with correspondingly high energy. The t_{2g} orbitals also split into d_{xz} and d_{yz} with lower energy but d_{xy} with correspondingly high energy. As shown in the figure, it is possible that the energy of d_{z^2} orbital drops below the energy of d_{xy}. This phenomenon is particularly observed in the square planar complexes of Co(II), Ni(II) and Cu(II).

Upon compression of these two trans-ligands, as expected, the exact opposite type of splitting will take place, which is shown in Figure 20.

6.1.2 The Molecular Orbital Theory

For an octahedral complex ML_6, let us consider that the ligands have only six sigma orbitals, viz, (σ_x, σ_{-x}, σ_y, σ_{-y}, σ_z, σ_{-z}) directed toward the metal ion. These six orbitals must be combined to obtain six ligand group orbitals in such a way that each of these orbitals is suitable for overlap with only one of the s, p or d orbitals of the metal, as shown in Figure 21.

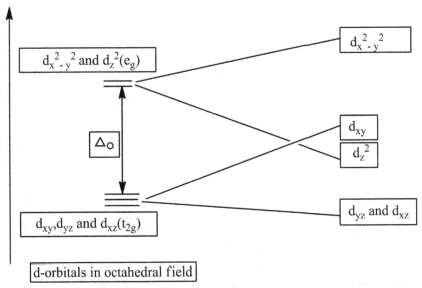

Figure 19 Further splitting of t_{2g} and e_g orbitals upon removing trans-ligands lying along the z axis.

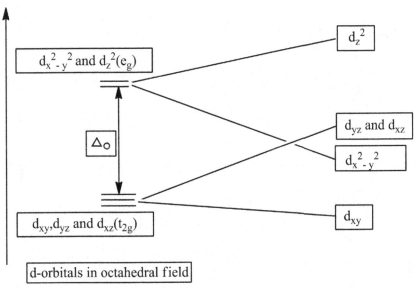

Figure 20 Further splitting of t_{2g} and e_g orbitals upon compressing trans-ligands lying along the z axis.

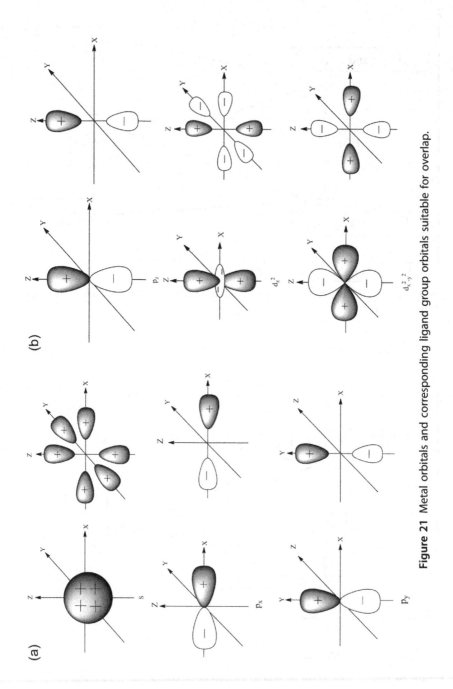

Figure 21 Metal orbitals and corresponding ligand group orbitals suitable for overlap.

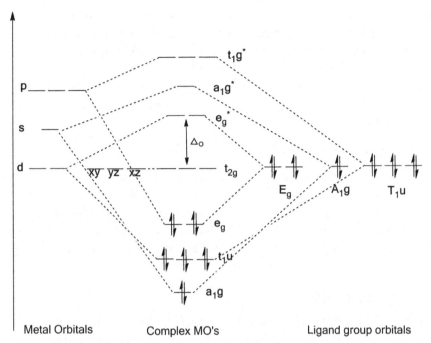

Figure 22 Molecular orbital diagram showing σ-bonding in an octahedral complex.

According to the MO theory, every overlap between ligand group orbitals and suitable metal orbitals will result into the formation of one bonding molecular orbital (BMO) and one antibonding molecular orbital (ABMO). The t_{2g} orbitals of the metals oriented in the plane cannot overlap with the ligand orbitals, and hence, they become nonbonding molecular orbitals. The e_g orbitals on linear combinations with ligand group orbitals give doubly degenerate e_g and e_g* molecular orbitals. The MO's derived from the p-orbitals are named as t_{1u} and $t_{1u}*$, respectively. These MO's are threefold degenerate, while the combination of s-orbital of the metal and ligand orbital results into the formation of molecular a_{1g} and $a_{1g}*$, respectively.

Figure 22 shows that the six electron pairs donated by ligands can be accommodated in the six sigma bonding molecular orbitals of the complex, resulting into the formation of six sigma bonds. From the MO diagram, it is clear that the MO theory also supports the splitting of d-orbitals into t_{2g} and e_g, as indicated by the LFT.

7. MAGNETISM

While the coordination chemistry was earlier the chemistry of compounds with coordinate covalent bonds, eventually it became the chemistry of transition elements with a wide variety of inorganic and organic ligands. The transition elements are characterized by incompletely filled d-orbitals. As a consequence, the transition elements and their compounds exhibit variable oxidation states, give coloured compounds, find application in catalysis and also have interesting magnetic properties [6]. An introductory study on the magnetic properties of transition metal complexes is provided in the following section.

7.1 Diamagnetism

The substances that, when placed in a magnetic field, decrease the intensity of the magnetic field than that in a vacuum are called diamagnetic substances, and this property is diamagnetism. The magnetic field tends to avoid such substances, and as such, diamagnetic substances are weakly repelled by the magnetic field. All of the chemical substances containing paired electrons show diamagnetism.

7.2 Paramagnetism

The substances that, when placed in a magnetic field, allow the magnetic field to pass through them as compared to a vacuum, are called paramagnetic substances, and the property of showing this behaviour is called paramagnetism. A paramagnetic substance tends to set itself with its length parallel to the magnetic field. Thus a paramagnetic substance is attracted into a magnetic field. The chemical substances containing unpaired electrons show paramagnetism.

7.3 Ferromagnetism and Antiferromagnetism

Ferromagnetism and antiferromagnetism are special classes of paramagnetism. A simple paramagnetism is observed in magnetically dilute systems, where the ions containing unpaired electrons are far apart from each other so that they can behave independently. Ferromagnetism and antiferromagnetism are observed in magnetically concentrated systems where the individual paramagnetic ions are very near, and they are affected by the magnetic moments of each other. In ferromagnetism, the magnetic moments point in the same direction. Due to this, the magnetic susceptibility of a substance increases to a great extent. Ferromagnetism is observed in transition metals and some of their compounds. In antiferromagnetism, the magnetic moments point in the opposite direction. Due to this, the magnetic susceptibility of a substance decreases to some extent. Antiferromagnetism is observed in the salts of ions like Mn^{+2}, Fe^{+3} and Gd^{+3}.

Magnetic susceptibility and magnetic moments are the important quantities in quantifying the magnetic properties of materials. The magnetic susceptibility, which is the response of a substance to induced magnetization, is represented in various forms like volume magnetic susceptibility (χ_v), gram magnetic susceptibility (χ_g) and molar magnetic susceptibility (χ_M). While the parent term, volume magnetic susceptibility is equated as $\chi_v = I/H$, which is the ratio of intensity of magnetization to the strength of the field. The other terms are derived from χ_v as $\chi_g = \chi_v/\rho$ (ρ represents the density) and $\chi_M = \chi_g \times M$ (M is the molecular weight of the substance).

The molar magnetic susceptibility χ_M is inversely proportional to T i.e. $\chi_M \, \alpha \, 1/T$ hence, $\chi_M = C/T$.

Where C is a **Curie's constant** and the equation is called Curie's equation. Ferromagnetism and antiferromagnetism are deviations from the Curie law. From Figure 23, it is seen that in both cases, the magnetic susceptibility changes suddenly at a temperature Tc – this temperature is called Curie temperature. For antiferromagnetism, the Curie temperature is also called Neel temperature.

The effective magnetic moment for a transition metal complex containing unpaired electrons has contributions from both orbital $[\vec{L}]$ as well as spin angular momentum $[\vec{S}]$, which can be represented as $\mu_{\text{eff}} = \sqrt{\vec{L}(\vec{L}+1) + 4\vec{S}(\vec{S}+1)}$ B.M. The spin-only magnetic moment μ_s can be calculated by the following formulae: $\mu_s = \sqrt{4S(S+1)}$ or $\sqrt{n(n+2)}$, which is illustrated below. The formula $\mu_s = \sqrt{4S(S+1)}$ can be used to calculate the spin only magnetic moment for a term symbol for a d^2 ion: ^3F using a general formula of a term symbol as ^{2S+1}L.

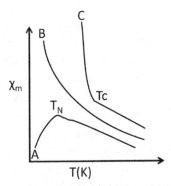

Figure 23 A plot of $\chi_M \rightarrow$ T showing different types of magnetic behaviour.

Thus, for ^3F, $2S + 1 = 3$ and hence $S = 1$.

Now, $\mu_s = \sqrt{4S(S+1)} = \sqrt{4 \times 1(1+1)} = \sqrt{4(2)} = \sqrt{8}$
$= 2.83$ B.M.

Similarly, the formula $\sqrt{n(n+2)}$, where 'n' represents the number of unpaired electrons, can be used for a d^2 ion in an octahedral field as

$\mu_s = \sqrt{n(n+2)} = \sqrt{2 \times (2+2)} = \sqrt{2(4)} = \sqrt{8} = 2.83$ B.M.

7.4 The Spectrochemical Series

The magnitude of splitting of d-orbitals; Δo or Δt is affected by the nature of the ligand. The electrons occupy the degenerate orbitals according to aufbau principle, Pauli's exclusion principle and Hund's rule of maximum spin multiplicity. But while filling the last electron in a set of degenerate orbitals, an option for putting the electron in the next higher orbital instead of pairing it in the same orbital comes into the picture. The actual action is decided by comparing the value of Δo or Δt, whichever is applicable with the pairing energy (P). While the pairing energy almost remains constant, the nature of the ligand affects the magnitude of splitting. Thus, the complexes of the same metal ion with different ligands sometimes have a different number of unpaired electrons. Experimental study of the spectra of a large number of complexes has made it possible to arrange the ligands in a series based on their capacity to cause the splitting of d-orbitals. This series is known as spectrochemical series.

$I^- > Br^- > Cl^- > F^- > OH^- > C_2O_4^{-2} > H_2O > \_NCS^- > py$
$> NH_3 > en > bpy > o\text{-phen} > NO_2^- > CN^-$

The initial members of the series cause lower values of the Δo and are known as weak field ligands, while the members toward the late end of the series cause greater splitting of the d-orbitals and are known as strong field ligands. The strong field ligands thus cause the electrons to pair in the degenerate orbitals at the cost of pairing energy. Hence, the resultant complexes will have low value of the total spin. Such complexes are known as low spin complexes. A counter-argument explains the definition of the high spin complexes are formed by weak field ligands like I^-, Br^-, Cl^-, F^- and OH^-. The spectrochemical series has limitations, as only the most common oxidation states of the metals are considered in making

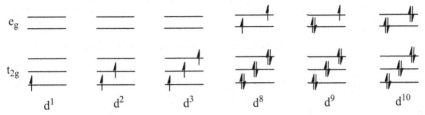

Figure 24 Only one possible arrangement of electrons in octahedral complexes containing 1, 2, 3, 8, 9 or 10 d-electrons.

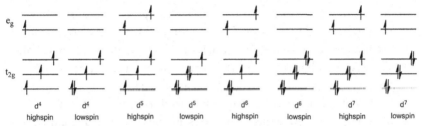

Figure 25 High-spin and low-spin arrangement of electrons in octahedral complexes containing 4, 5, 6 or 7 d-electrons.

the spectrochemical series, and even with the common oxidation states of some metals, a small variation in the sequence of members is found sometimes.

Since the number of unpaired electrons in high-spin and low-spin states are different, the magnetic moments in the corresponding complexes are also expected to be different.

As shown in Figure 24, the number of unpaired electrons in the presence of a strong field and weak ligands remain the same when we have 1, 2, 3, 8, 9 or 10 d-electrons in octahedral complexes.

As shown in Figure 25, the number of unpaired electrons in high-spin and low-spin states are different for 4, 5, 6 or 7 d-electrons in octahedral complexes.

8. CATALYTIC PROPERTIES

The transition metals and their compounds are well-known for their catalytic properties. The works related to catalytic applications involving metallic compounds have been awarded about 15 Nobel prizes between 1901 and 2014. While the idea of catalysis was originally conceived by Berzelius, it was

Wilhelm Ostwald who was first recognized for a Nobel prize on catalysis in the field of chemistry. Sabatier, a Nobel laureate (1912), found that the introduction of a trace amount of nickel to the reaction mixture catalyzes the hydrogenation of many organic compounds. The Sabatier principle [7] for catalysis emphasizes on the strength of the interaction between substrate and catalyst. He mentioned that in order to have a fruitful catalysis of a reaction, this interaction should neither be too strong or too weak; it should be just right. Other Nobel-winning achievements in the field of catalysis include the Haber-Bosch process (1918) for the catalytic manufacture of ammonia and catalytic cracking using metal oxides. Metallocenes, a class of organometallic compounds, found application in catalyzing the polymerization of alkenes containing terminal double bond. These compounds are very popular in the polymer industry as Ziegler–Natta catalysts [8]. Wilkinson's catalyst, chlorotris(triphenylphosphine)rhodium(I), is a prominent transition metal complex employed as a catalyst in the hydrogenation of alkenes [9].

The twenty-first century transition metal complex catalysts include Grubb's catalysts for metathesis reactions in olefins and Palladium complexes used for coupling reactions in organic chemistry.

A series of transition metal carbene complexes, known as Grubb's [10,11] first generation and second generation catalysts and Hoveyda–Grubb's first and second generation catalysts, afford metathesis between two olefins. The metathesis reactions refer to the exchange of the cationic and anionic partners and may be represented by the following general equations, as in Figure 26.

The mechanisms of these reactions involve a replacement of phosphine ligand by an olefin and a subsequent metathesis reaction on the metal complex itself. A variety of Ruthenium carbene complexes used as catalysts are shown in Figures 27 and 28.

A very important reaction, known as a Suzuki coupling reaction, which involves the coupling of organic groups by boronic acid and a halide, is known to be catalyzed by Pd(0) complex [12–14]. The reaction may be represented as shown in Figure 29.

The mechanism of this reaction involves a shuttle between 0 and +2 oxidation states of Palladium. The mechanism of this reaction can be explained by focussing on the reactions occurring on the palladium catalyst.

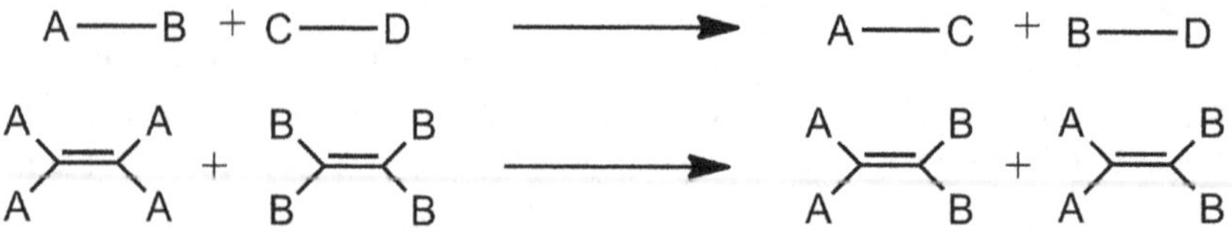

Figure 26 Simplified representations for metathesis reactions.

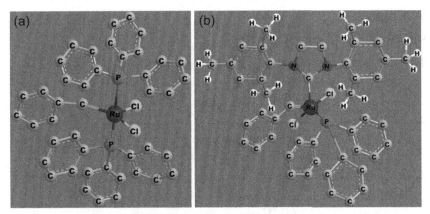

Figure 27 Grubb's first and second generation catalysts.

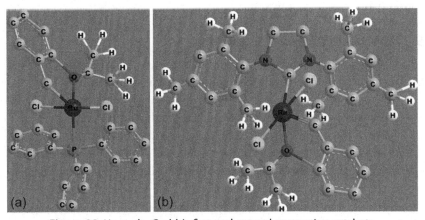

Figure 28 Hoveyda–Grubb's first and second generation catalysts.

$$R_1\!-\!\!-BY_2 + R_2\!-\!\!-X \xrightarrow[\text{OH}^-]{\text{Pd(0) complex catalyst}} R_1\!-\!\!-R_2$$

Figure 29 Suzuki coupling reaction.

In the first step, an oxidative addition of palladium to the halide gives an organopalladium species. A subsequent reaction with a base and transmetallation followed by reductive elimination of the product gives back the palladium catalyst, completing the catalytic cycle. The complex involved in the catalytic cycle bears the general formula: $[\text{ArPd(OR)L}_2]$.

9. EXERCISES

9.1 Multiple Choice Questions

1. What is the spin multiplicity of the electronic state '^3P'
   (a) 1   (b) 1/2
   (c) 3   (d) 2

2. Antiferromagnetism is a type of ——————
   (a) Magnet   (b) Diamagnetism
   (c) Paramagnetismt (d) Ferromagnetism

3. CN^- is a ———————————— field ligand.
   (a) Bidentate   (b) Octahedral
   (c) Strong   (d) Weak

4. What is the value of 'L' for ^1D?
   (a) 4   (b) 1
   (c) 2   (d) None

9.2 Short/Long Answer Questions

1. Giving suitable diagrams, describe the splitting of d-orbitals upon the introduction of an octahedral and tetrahedral ligand field.
2. What is centre of gravity rule? How does it apply to the splitting of the d-orbitals in the octahedral and tetrahedral ligand fields?
3. Prepare a diagram that traces how the d-orbital splitting pattern changes as an octahedral complex is altered via tetragonal distortions.
4. Explain how the MO theory supports the splitting of d-orbitals as indicated by the crystal field theory.
5. By using orbital splitting diagrams, show which d^n electron configurations are capable of giving both low-spin and high-spin configurations in an octahedral ligand field.
6. Give an account of the magnetic properties of transition metal complexes.
7. Calculate the spin-only magnetic moments in d^4, d^5, d^6 and d^7 complexes in both low-spin and high-spin configurations in an octahedral ligand field.
8. What is a spectrochemical series? What are its limitations?
9. Calculate the LFSE in the units of Δo for the following high-spin ions in octahedral fields Fe^{+2}, Mn^{+2} and Co^{+2}.
10. Predict the magnetic properties and LFSE for each of the following $[Fe(CN)_6]^{-3}$, $[Co(NH_3)_6]^{+3}$, $[Co(Cl)_4]^{-2}$ and $[Fe(H_2O)_6]^{-3}$.

SUGGESTED FURTHER READINGS

The topics discussed in this chapter are a part of standard graduate curriculum. A majority of the textbooks with titles related to basic coordination chemistry can act as a source of further reading. Moreover, there are several web resources useful for further learning. Some of them are listed below.

http://chemed.chem.purdue.edu/genchem/topicreview/bp/ch12/complex.php#werner

http://www.britannica.com/science/coordination-compound

http://nptel.ac.in/courses/104103069/20

http://www.chem.iitb.ac.in/people/Faculty/prof/pdfs/L5.pdf

REFERENCES

Attempts have been made to include some illustrations from the books and research articles listed below.

[1] Smith HM. Alfred Werner: 1866–1919. In: Smith HM, editor. Torchbearers of chemistry. Academic Press; 1949. p. 253.

[2] Connelly NG, Royal Society of Chemistry (Great Britain), International Union of Pure and Applied Chemistry. Nomenclature of inorganic chemistry. IUPAC recommendations 2005, vol. xii. Cambridge, UK: Royal Society of Chemistry Publishing/IUPAC; 2005. 366 p.

[3] House JE, House KA. Chapter 19–Structure and bonding in coordination compounds. In: House JE, House KA, editors. Descriptive inorganic chemistry. 2nd ed. Amsterdam: Academic Press; 2010. p. 441–78.

[4] Burns G. Chapter 8–Crystal field theory (and atomic physics). In: Burns G, editor. Introduction to group theory with applications. Academic Press; 1977. p. 156–98.

[5] Graddon DP. Chapter II–Modern theories of co-ordination chemistry. In: Graddon DP, editor. An introduction to co-ordination chemistry. 2nd ed. Pergamon; 1968. p. 24–54.

[6] Earnshaw A. V–further topics. In: Earnshaw A, editor. Introduction to magneto-chemistry. Academic Press; 1968. p. 70–83.

[7] Medford AJ, et al. From the Sabatier principle to a predictive theory of transition-metal heterogeneous catalysis. J Catal 2015;328:36–42.

[8] Coates GW, Waymouth RM. 12.1–Transition metals in polymer synthesis: Ziegler–Natta reaction. In: Wilkinson EWAGAS, editor. Comprehensive organometallic chemistry II. Oxford: Elsevier; 1995. p. 1193–208.

[9] Pettinari C, Marchetti F, Martini D. Metal complexes as hydrogenation catalysts. In: Meyer JAMJ, editor. Comprehensive coordination chemistry II. Oxford: Pergamon; 2003. p. 75–139.

[10] Penczek S, Grubbs RH. 4.01–Introduction. In: Möller KM, editor. Polymer science: a comprehensive reference. Amsterdam: Elsevier; 2012. p. 1–3.

[11] Grubbs RH, Pine SH. 9.3–Alkene metathesis and related reactions. In: Fleming BMT, editor. Comprehensive organic synthesis. Oxford: Pergamon; 1991. p. 1115–27.

[12] Moriya T, et al. A new facile synthesis of 2-substituted 1,3-butadiene derivatives via palladium-catalyzed cross-coupling reaction of 2,3-alkadienyl carbonates with organoboron compounds. Tetrahedron 1994;50(27):7961–8.

[13] Miyaura N, et al. Palladium-catalyzed cross-coupling reactions of B-alkyl-9-BBN or trialkylboranes with aryl and 1-alkenyl halides. Tetrahedron Lett 1986;27(52):6369–72.

[14] Miyaura N, Yano T, Suzuki A. The palladium-catalyzed cross-coupling reaction of 1-alkenylboranes with allylic or benzylic bromides. Convenient syntheses of 1,4-alkadienes and allybenzenes from alkynes via hydroboration. Tetrahedron Lett 1980;21(30):2865–8.

CHAPTER 2

Basic Concepts of Symmetry and Group Theory

Contents

Essentials of Coordination Chemistry
http://dx.doi.org/10.1016/B978-0-12-803895-6.00002-1

1. INTRODUCTION

The symmetry of molecules and solids is a very powerful tool for developing an understanding of bonding and physical properties of various chemical entities. At the same time, it allows a learner to quantify the beauty of molecules with different shapes. The knowledge of symmetry is useful in predicting the nature of molecular orbitals. It also helps in predicting whether electronic and vibration spectroscopic transitions can be observed. The presence and absence of elements of symmetry in molecules leads a chemist to predict the possible optical isomerism in molecules. Thus, symmetry becomes an important tool in the hand of a theoretical and inorganic, as well as an organic, chemist.

2. SYMMETRY OPERATIONS AND ELEMENTS OF SYMMETRY

2.1 Rotation

Rotation is an operation referring to a circular movement of an object about an axis passing through the centre of an object.

One full rotation corresponds to a movement of $360°$ about an axis, and it is the same as a rotation of $0°$ (practically no rotation). Thus, after a rotation of $360°$, the appearance of the original object and the resultant object always remain indistinguishable.

The axis of rotation is named as C_n, where n represents the number of times indistinguishable appearance was obtained upon rotation of $360°$. If $n = 1$, the rotation is through an angle of $360°$, which means no change to the molecule. Thus C_1 means **identity operation** (E), which leaves the molecule unmoved. Doing no change to the object is called identity operation. This operation is needed for mathematical completeness. Every molecule has at least this symmetry element.

An object can have more than one axis of rotation about which an indistinguishable appearance of the object is obtained more than once upon rotation of $360°$.

The axis with highest value of n is called the **principal axis**.

Axes of rotation in a few hypothetical molecules with one- and two-dimensional geometries are discussed in the following section.

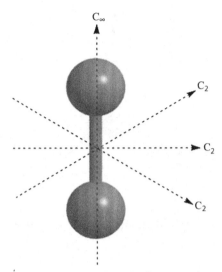

Figure 1 Infinite fold rotational axis (C_∞) and infinite number of C_2 axes in a linear molecule.

2.1.1 Linear Molecules

A diatomic molecule rotated along the axis, as shown in Figure 1, becomes indistinguishable infinite times (remains indistinguishable) during the rotation of 360°, hence the axis is named as C_∞, which is the principal axis. There are infinite numbers of axes perpendicular to C_∞ and passing through the centre of the molecule, which yields the original appearance twice during a rotation of 360°. All these axes are named as C_2, C_2', C_2'' and so on.

2.1.2 Angular Molecules

A triatomic angular molecule of the type A_3 or AB_2 becomes indistinguishable twice during the rotation of 360°, hence the axis is named as C_2, as shown in Figure 2.

2.1.3 Triangular Molecules

A triatomic triangular molecule of the type AB_2 becomes indistinguishable twice during the rotation of 360°, hence the axis is named C_2. It has three such axes passing through each of the atoms and coincides with the centre of the molecule. Such molecules have a C_3 axis (principal axis)

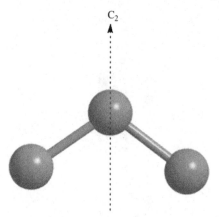

Figure 2 The C_2 axis of rotation in an angular molecule.

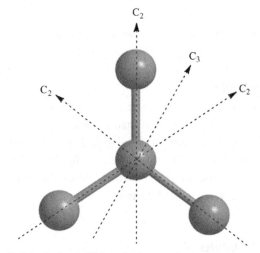

Figure 3 Principal axis (C_3) and $3C_2$ axes in a triangular molecule.

perpendicular to the plane of the triangle and passing through the centre of the triangle, as shown in Figure 3.

2.1.4 Square Planar Molecules

A square planar molecule of the type MA_4 have a principal axis C_4 perpendicular to the plane of the square and passing through the centre of the square, as shown in Figure 4. Additionally, there are two C_2 axes that include two opposite atoms and the centre of the molecule. The other two C_2 axes pass between the two adjacent atoms and also include the centre of the molecule. In order to keep the figure simple, only one C_2 axis of each type is shown.

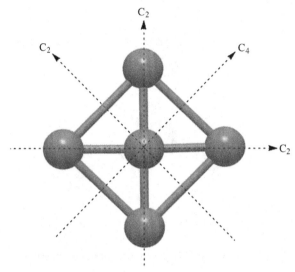

Figure 4 Principal axis (C_4) and C_2 axes in a square planar molecule.

The molecules with coordination number four and higher include the three-dimensional geometries. Axes of rotation in a few such hypothetical molecules are discussed in the following section.

2.1.5 Tetrahedral Molecules

A tetrahedral molecule of the type MA_4 has four principal axes (C_3) passing through each of the terminal atoms and the central atom (one of these is shown in Figure 5). Additionally, there are three C_2 axes passing between two terminal atoms and the centre of the molecule (one of these is shown in Figure 5).

2.1.6 Trigonal Bipyramidal Molecules

In a trigonal bipyramidal molecule of the type MA_5, the two axial A's form a straight line with the centre, while the remaining three A's (separated by an angle of $120°$) form a triangle at the equator. The line joining the two axial A's is the principal axis C_3, while three axes passing through each of the equatorial atoms and the central atom are C_2 axes (one of these is shown in Figure 6).

Consider a trigonal bipyramidal $[CuCl_5]^{3-}$ ion, as shown in Figure 7.

Here, rotation about a C_3 axis in clockwise and anticlockwise directions restores the identity of the molecule. However, both these operations yield the molecules with the positions of chlorines that are different from the

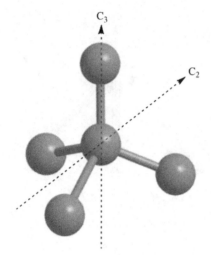

Figure 5 Principal axis (C_3) and C_2 axis in a tetrahedral molecule of type MA$_4$.

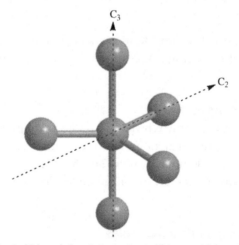

Figure 6 Principal axis (C_3) and C_2 axis in a trigonal bipyramidal molecule of type MA$_5$.

original one. It can also be seen from the figure that C_3 performed twice (i.e. $C_3 \times C_3 = C_3^2$) gives the same result as that of an anticlockwise rotation. Also $C_3 \times C_3 \times C_3 = C_3^3 = E$.

2.1.7 Octahedral Molecules

An octahedral molecule of the type MA$_6$ clearly has three C_4 axes passing through opposite atoms, which are the principal axes. It also has

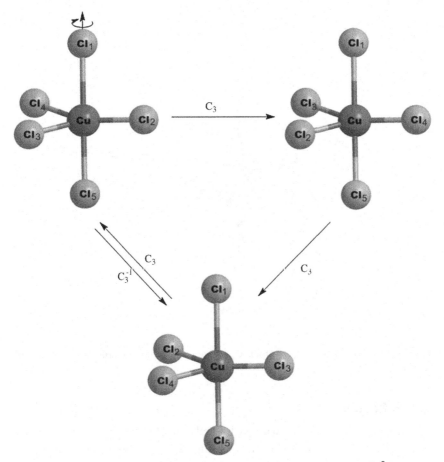

Figure 7 Rotation operations around the principal axis (C_3) in a $[CuCl_5]^{3-}$ ion.

four C_3 axes of rotation passing through the centroid of the opposite triangles and including the centre of the molecule. Moreover, six lines passing between the adjacent atoms account for the six C_2 axes in a regular octahedron. A sample axis of each of the above mentioned type is shown in Figure 8.

As shown in Figure 9, a symmetrically substituted octahedral complex ion $[Fe(CN)_6]^{3-}$ has a four-fold axis of symmetry passing through $N-C_1-Fe-C_6-N$. Four different positions of the ligands restoring the original appearance are obtained by carrying out the C_4 rotation successively. Here also $C_4 \times C_4 \times C_4 \times C_4 = C_4^4 = E$.

This observations leads us to generalize that $C_n^n = E$.

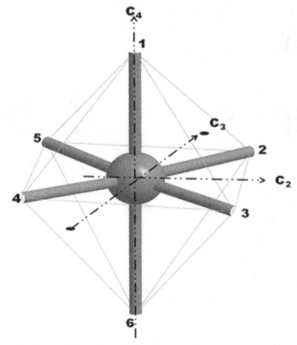

Figure 8 Sample axes of the types C_4, C_3 and C_2 in an octahedral molecule of type MA_6.

2.2 Reflection

If a molecule remains unchanged by reflection in a plane that is passing through the centre of the molecule, the plane is said to be a plane of symmetry.

The plane of reflection is denoted by symbol σ (sigma).

Remember that double reflection in the same plane always gives identity (i.e. $\sigma \times \sigma = \sigma^2 = E$).

If the plane of reflection is perpendicular to the principal axis, the symbol is σ_h. A σ_h (the molecular plane) perpendicular to the principal axis C_4 in a square planar $[Ni(CN)_4]^{-2}$ ion is shown in Figure 10.

If the plane of reflection contains the principal axis, the symbol is σ_v. Angular molecule water has two such planes: the molecular plane and the plane perpendicular to the molecular plane, as shown in Figure 11.

A special class of σ_v plane, σ_d (dihedral), is found in some molecules. σ_d is found in molecules having C_2 axes at the right angles to the principal axis. If the plane of reflection bisect the angles between two adjacent C_2 axes and

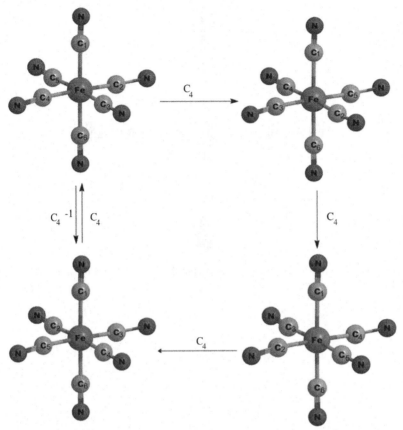

Figure 9 Rotation operations around the principal axis (C_4) in $[Fe(CN)_6]^{3-}$ ion.

pass through the principal axis, they are called σ_d planes. The σ_d in a square planar complex $[Ni(CN)_4]^{-2}$ ion is shown in Figure 12.

2.3 Improper Rotation (Rotation–Reflection)

A rotation by $2\Pi/n$ about an axis followed by reflection in a plane perpendicular to the axis of rotation (σ_h) is called improper rotation (Rotation–Reflection).

The axis is called the Rotation–Reflection axis and the symbol is S_n.

Remember that $S_n = C_n \times \sigma_h = \sigma_h \times C_n$, which means that the sequence of performance of the operations does not affect the final result.

Also remember that when n is even, $S_n^n = C_n^n \times \sigma_h^n = E \times E = E$, which means that the n-fold repetition of the improper rotation returns the

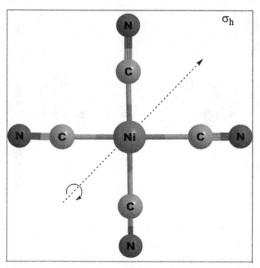

Figure 10 The molecular plane (σ_h) perpendicular to the principal axis C_4 in $[Ni(CN)_4]^{-2}$ ion.

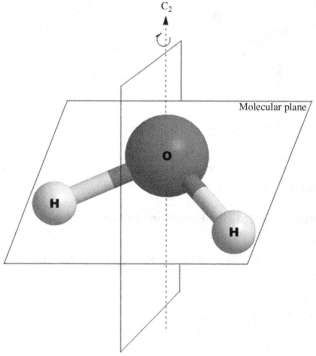

Figure 11 The molecular plane (σ_v) and the plane perpendicular to molecular plane (also σ_v) in a water molecule.

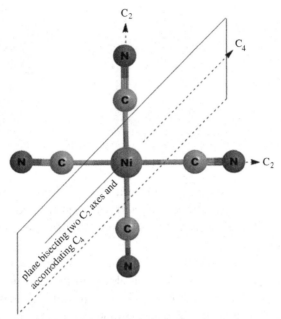

Figure 12 Dihedral plane (σ_d) in $[Ni(CN)_4]^{-2}$ ion.

molecule to its initial position. And when n is odd, $S_n^n = C_n^n \times \sigma_h^n = E \times \sigma_h = \sigma_h$, which means that the n-fold repetition of the improper rotation is equivalent to a simple reflection in a plane perpendicular to the axis.

From Figure 13, it is seen that a phosphazene trimer $(PNCl_2)_3$ has an S_3 axis of symmetry. The C_3 axis of rotation is perpendicular to the equilateral triangle obtained by joining three phosphorous atoms. The two chlorine atoms bound to each phosphorous atoms are above and below the plane of the triangle. A rotation of $120°$ followed by reflection in the plane perpendicular to this axis retains the appearance of the molecule with the new positions of the chlorine atoms. A few more operations on the above mentioned molecule can also lead us to derive the following relations.

$$S_3^2 = S_3 \times S_3 = C_3^2, S_3^3 = \sigma_h, S_3^4 = C_3, S_3^5 = C_3^2 \times \sigma_h, S_3^6 = E$$

2.4 Inversion

A rotation by $180°$ about an axis followed by reflection in a plane perpendicular to the axis of rotation (σ_h) is called inversion.

Inversion operation is denoted by i where $i = S_2 = C_2 \times \sigma_h$.

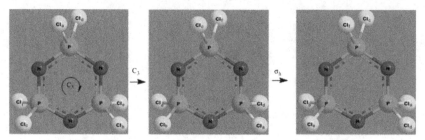

Figure 13 Improper rotation (S_3) in a phosphazene trimer.

Practically, an inversion operation can be carried out by interchanging the opposite positions in a molecule, and if the appearance remains indistinguishable from the original one, one can say that the molecule has a centre of symmetry.

From Figure 14, it is seen that the *trans*-isomer of dichloroethylene has a centre of symmetry while this centre is absent in the *cis*-isomer.

From the preceding discussion, we have seen that molecules with different geometries can have the same appearance subsequent to operations like rotation around various axes, reflections and improper rotation, as well as inversion. Any of these operations on a body after which the appearance of the object remains indistinguishable from the original one is called symmetry operation. Each of the molecules can have more than one axis, as well as planes of symmetry. These axes and planes are known as the elements of symmetry.

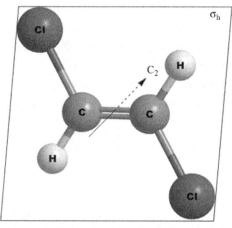

Figure 14 The rotational axis (C_2) and plane of reflection (σ_h) for inversion of *trans*-dichloroethylene.

3. POINT GROUPS

3.1 Symmetry Group

A collection of the elements of symmetry of a molecule is called a symmetry group.

3.2 Point Group

If, in a molecule, at least one point remains fixed when any of the operations of the symmetry group is carried out, then the symmetry group is called the point group.

The elements of a point group obey the following rules:

1. Each group must contain the identity element E.
2. The elements of a group can be multiplied by one another. The product of two operations is the result of applying them in succession. Hence, the product of any two elements of a group is also an element of that group.
3. The elements of a group follow the associative rule of multiplication, i.e. $A(BC) = (AB)C$, but are not necessarily commutative, i.e. $AB = BA$. If all the elements of the group follow commutative rule, the group is called Abelian.
4. For every element A in a group, there is an element in the group A^{-1} (A inverse), which is the reverse operation of A. Thus $AA^{-1} = A^{-1}A$.

3.3 Some Point Groups

3.3.1 C_n, C_s and C_i Groups

The molecules containing only the C_n axis as the element of symmetry operation other than E are classified in these groups. They contain the total n number of symmetry elements. These elements are C_n, C_n^2, $C_n^3 \ldots C_n^n$, where $C_n^n = E$.

3.3.1.1 C_1

C_1 has only one symmetry element: E. A highly unsymmetrical tetrahedral complex of the type M_{abcd}, shown in Figure 15a, belongs to the C_1 point group.

3.3.1.2 C_s

Tetrahedral complexes and molecules, shown in Figure 1, with two or three identical ligands (M_{a_2bc} and M_{a_3b}) have only one element of symmetry in addition to E. This element is a plane between the identical ligands. Such entities are classified into a C_s point group. The C_s group is

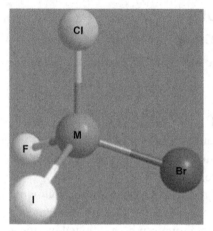

Figure 15a Tetrahedral complex (M_{abcd}) with only one element of symmetry E.

represented by species such as CH_2BrCl, CH_3OH and any other tetrahedral complexes with the general formula $M_{a_2bc}M_{a_3b}$.

3.3.1.3 C_i

Figure 15b represents a dinuclear complex formed by a pair of tetrahedral complexes M_{abc} and a metallic bond. Such a molecule has only an identity element of symmetry. However, as all the identical ligands are placed exactly opposite to each other, an indistinguishable appearance can be obtained by interchanging the opposite positions, which means that it possesses the centre of inversion i. Hence, this complex is an example of the point group C_i. The meso form of the organic compound, CHClBr—CHClBr, is another example of the point group C_i.

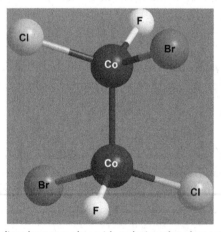

Figure 15b A dinuclear complex with only i as the element of symmetry.

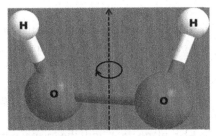

Figure 16 C_2 axis of rotation in H_2O_2.

3.3.1.4 C_2

C_2 has two symmetry elements: C_2 and $C_2^2 = E$. Hydrogen peroxide (H_2O_2) with an 'open-book' type of geometry, as shown in Figure 16, is an illustration of this point group.

3.3.1.5 C_3

C_3 has three symmetry elements: C_3, C_3^2 and $C_3^3 = E$. A molecule $B(OCH_3)_3$, shown in Figure 17, represents the C_3 point group, and the rotation around this axis is demonstrated by a circular path in the figure. Octahedral chelates of type $M(AB)_3$, illustrated by trisethanolamine Chromium (III) (as shown in Figure 17), also belong to the C_3 point group.

3.3.1.6 C_6

C_6 has six symmetry elements: C_6, $C_6^2 = C_3$, $C_6^3 = C_2$, $C_6^4 = C_3^2$, C_6^5 and $C_6^6 = E$. All the elements in these groups commute with each other, and hence, all these groups are Abelian.

(a)   (b)

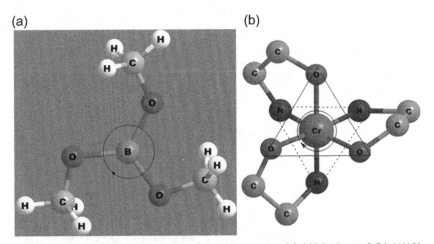

Figure 17 Representative examples of C_3 point group. (a) $B(OCH_3)_3$, and (b) $M(AB)_3$ chelate.

3.3.2 C_{nv} Groups

The molecules containing one or more planes (vertical σ_v) of symmetry, in addition to the C_n axis, are classified in C_{nv} groups.

3.3.2.1 C_{2v}

From the previous discussion, we now know that an angular molecule like water (OH_2) has a C_2 as a principal axis. It also has two vertical planes of symmetry: the molecular plane (σ_v) and the plane perpendicular to the molecular plane (σ_v'), as shown in Figure 18.

The relation, $C_2 \times \sigma_v = \sigma_v'$ is explained in Figure 18.

Thus the point group C_{2v} contains elements E, C_2, σ_v and σ_v'.

In order to check whether the elements of the group commute with each other or not, a multiplication table shown in Table 1 is required to be created.

From the multiplication table, it is clear that the net result is independent of the sequence of the operations (i.e. $AB = BA$). Thus the group C_{2v} is an Abelian group.

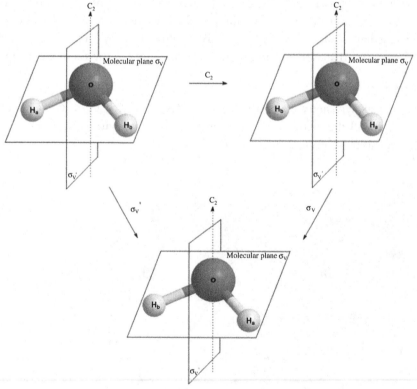

Figure 18 The relation $C_2 \times \sigma_v = \sigma_v'$ for a water (OH_2) molecule.

Table 1 Multiplication table for C_{2v} point group
First Operation →

Second Operation ↓	E	C_2	σ_v	σ_v'
E	E	C_2	σ_v	σ_v'
C_2	C_2	E	σ_v'	σ_v
σ_v	σ_v	σ_v'	E	C_2
σ_v'	σ_v'	σ_v	C_2	E

3.3.2.2 C_{3v}

The pyramidal ligands, such as NH_3 and PH_3, have a C_3 axis and three σ_v planes (σ_{va}, σ_{vb} and σ_{vc}) that include the hydrogen atoms labelled a, b and c, respectively, along with the nitrogen atom of ammonia, as shown in Figure 19.

The point group C_{3v} contains E, two rotation operators C_3 and C_3^2 and three vertical planes of symmetry σ_{va}, σ_{vb} and σ_{vc}. Figure 19 shows that $C_3 \times \sigma_{va} = \sigma_{vb}$ and $\sigma_{va} \times C_3 = \sigma_{vc}$, meaning that $C_3 \times \sigma_{va} \neq \sigma_{va} \times C_3$ (which is not a commutator relationship), hence the group C_{3v} is non-Abelian. While performing successive operations, one must keep in mind that σ_{va} is the plane containing H_a and N as per the starting arrangement. The results of the various symmetry operations performed for a C_{3v} point group are summarized in Table 2.

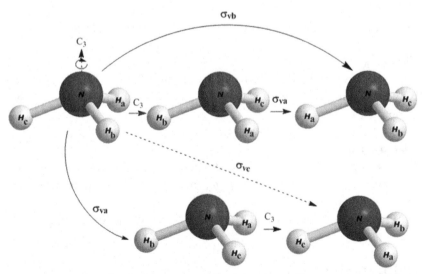

Figure 19 The non-Abelian nature of C_{3v} point group ($C_3 \times \sigma_{va} \neq \sigma_{va} \times C_3$).

Table 2 Multiplication table for C_{3v}
First Operation $\rightarrow$

Second Operation $\downarrow$	E	C_3	C_3^2	σ_{va}	σ_{vb}	σ_{vc}
E	E	C_3	C_3^2	σ_{va}	σ_{vb}	σ_{vc}
C_3	C_3	C_3^2	E	σ_{vc}	σ_{va}	σ_{vb}
C_3^2	C_3^2	E	C_3	σ_{vb}	σ_{vc}	σ_{va}
σ_{va}	σ_{va}	σ_{vb}	σ_{vc}	E	C_3	C_3^2
σ_{vb}	σ_{vb}	σ_{vc}	σ_{va}	C_3^2	E	C_3
σ_{vc}	σ_{vc}	σ_{va}	σ_{vb}	C_3	C_3^2	E

3.3.2.3 $C_{\infty v}$

They have an infinite fold rotation axis, as shown in Figure 1. A heteroatomic molecule (AB) or a linear complex of type M_{ab} belongs to the $C_{\infty v}$ group.

3.3.3 C_{nh} Groups

These groups contain n rotation operators and n-operators in which each rotation is multiplied by σ_h.

$$C_n, C_n^2, C_n^3 \ldots C_n^n = E, S_n^n = C_n^n \times \sigma_h^n, C_n^2 \times \sigma_h, C_n^3 \times \sigma_h \ldots C_n^n \times \sigma_h$$

C_{nh} groups contain a total $2n$ number of elements, and they are Abelian. C_{nh} groups have the centre of inversion i when n is even, while the centre of inversion is absent for odd values of n.

3.3.3.1 C_{1h}

The point group C_{1h} has two elements of symmetry: E and σ_h.

3.3.3.2 C_{2h}

The point group C_{2h} has four elements of symmetry: E, C_2, σ_h and i. An example of this point group is provided by *trans*-dichloroethylene, shown in Figure 14.

3.3.4 S_n Groups

When n is even, $S_n^n = C_n^n = E$, hence there will be n symmetry operators.

When n is odd, $S_n^n = C_n^n \times \sigma_h^n = \sigma_h$ and $S_2^{2n} = E$, hence there will be $2n$ symmetry operators.

When n is odd as S_n has both σ_h and $C_n^x = S_n^{2x}$, it is the same as C_{nh} ($n = $ odd).

3.3.5 D_n Groups

They have n two-fold axes perpendicular to the principal n-fold axis. There is no plane of symmetry.

3.3.5.1 D_3
The chelate trisethylenediamine Co(III) has D_3 symmetry. Figure 20 shows principal axis C_3 and three C_2 axes perpendicular to it.

3.3.6 D_{nd} Groups
An addition of σ_d operation to D_n gives the D_{nd} group. The allene molecule (CH_2=C=CH_2) shown in Figure 21 has a dihedral plane passing through the two hydrogen atoms in addition to the C_2 axis. Thus it belongs to the point group D_{2d}.

3.3.7 D_{nh} Groups
An addition of σ_h operation to D_n gives D_{nh}.

3.3.7.1 D_{2h} (Vh)
Pyrazine has the elements of symmetry for the point group D_{2h}, as shown in Figure 22. It has three mutually perpendicular two-fold axes. It also has three mutually perpendicular planes of symmetry. As a result, pyrazine also has a centre of symmetry.

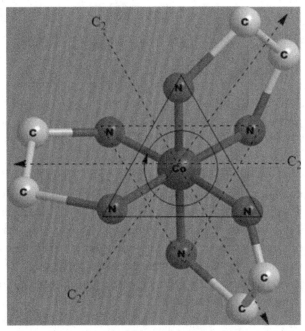

Figure 20 Three C_2 axes perpendicular to the principal axis C_3 in trisethylenediamine Co(III) chelate.

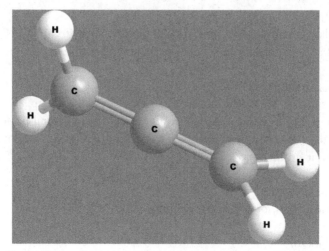

Figure 21 An allene molecule belonging to the point group D_{2d}.

(a)   (b)

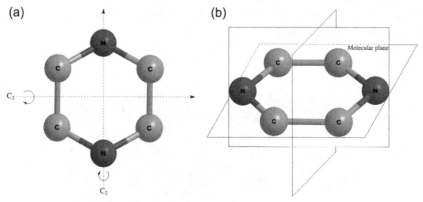

Figure 22 Pyrazine molecule. (a) Three mutually perpendicular C_2 axes, and (b) three mutually perpendicular σ_h.

3.3.7.2 D_{6h}

Benzene belongs to the point group D_{6h}. The symmetry elements, as shown in Figure 23, include the principal axis C_6 and six C_2 axes perpendicular to the principal axis. Out of these six axes, three pass through the opposite corners of the hexagon, whereas the remaining three pass through the midpoints of the opposite sides of the hexagon. The molecular plane acts as σ_h, imparting D_{6h} symmetry point group to the molecule. In all, D_6 has 12 symmetry elements.

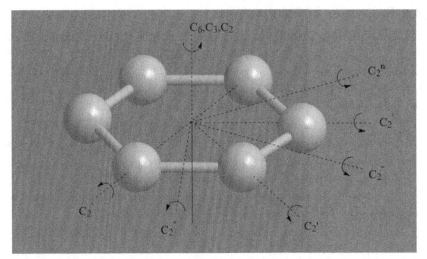

C_6, C_3, C_2

C_2^n

C_2

C_2''

C_2

C_2''

C_2'

Figure 23 Six C_2 axes perpendicular to the principal axis C_6 and a σ_h plane in benzene.

3.3.8 Cubic Point Groups

3.3.8.1 T

The group T has four C_3 axes and three C_2 axes of symmetry, as shown in Figure 5.

3.3.8.2 T_d

Addition of σ_d to T gives the T_d point group.

3.3.8.3 T_i (T_h)

Addition of the centre of symmetry to the T group gives the T_i or T_h point group.

3.3.8.4 O

There are three C_4, four C_3 and six C_2 axes for an O group, as shown in Figure 8. In all, there are 21 elements of symmetry for a point group O, which may be identified out of interest.

3.3.8.5 O_h

Addition of the centre of symmetry to the O group gives O_h. The point group O_h has 48 elements of symmetry in all, which may be identified out of interest.

3.4 Properties of Point Groups

Consider a group G with g number of elements, as shown in Figure 24. If $H_1, H_2...H_h$ are the elements of the group G that obey all the four rules of the point group, then these elements $H_1, H_2...H_h$ form a subgroup, say 'H'. Let F be a member of G but not H. Also, for $F \times H_1$, $F \times H_2...F \times H_h$, the products may be the members of G but not H.

If $F \times H_i = H_j$ were one of the members of H, then $F = H_j \times H_i^{-1}$, which means that F is obtained by operation between two members of sub group H. Hence, F must be a member of subgroup H.

Thus there will be $2h$ such members in the group G.

If $2h < g$, it will be possible to find another element D, which is neither present among $H_1, H_2...H_h$ nor $F \times H_1$, $F \times H_2...F \times H_h$!

Repeating the previous argument, we can get h new elements, viz, $D \times H_1$, $D \times H_2...D \times H_h$.

Finally we will get $k \times h = g$, where k is an integer.

'A group of six elements can have subgroups of order (number of members) one or two or three only'. The above statement can be explained by following arguments:

We have already seen that $k \times h = g$, where k is an integer, g is the total number of members in the group, and h is the number of members in the subgroup (order).

Hence, in this case, $k \times h = 6$

This number 6 can be obtained in only three ways: $6 \times 1 = 6$, $3 \times 2 = 6$ and $2 \times 3 = 6$ (i.e. by putting $h = 1$, 2 or 3). Thus the above statement is justified.

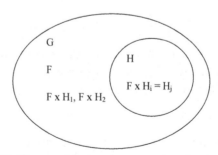

Figure 24 Diagram showing the properties of point groups.

3.5 Conjugate Elements

Elements A and B of a group are said to be conjugate if $A = X^{-1}BX$, where X is another element of the same group.

The transformation $A = X^{-1}BX$ is called similarity transformation.

Now if A and B are conjugate elements, we have $A = X^{-1}BX$

Multiplying the left end of the equation by X and the right end of the equation by X^{-1} on both sides, we get $XAX^{-1} = XX^{-1}BX\,X^{-1}$, which can also be written as $XAX^{-1} = XX^{-1}BXX^{-1}$.

Since $XX^{-1} = 1$, the relation becomes $XAX^{-1} = B$, which shows that the conjugate property is mutual.

3.6 Class

A complete set of elements, such as A, B, C...etc. of a group, which are mutually conjugate, is called a class. Every class can be completely determined from any one of its elements. The group C_{3v} has six operations, and they are divided into the following three classes:

1. E
2. C_3 and C_3^2
3. σ_{va}, σ_{vb} and σ_{vc}

We can show that C_3 and C_3^2 are conjugates of each other as follows:

$$E^{-1} \times C_3 \times E = E \times C_3 \times E = C_3$$
$$C_3^{-1} \times C_3 \times C_3 = C_3^2 \times C_3 \times C_3 = C_3^2 \times C_3^2$$
$$\left(C_3^2\right)^{-1} \times C_3 \times C_3^2 = C_3 \times C_3^3 \times C_3 = C_3^2 \times E = C_3$$
$$\sigma_{va}^{-1} \times C_3 \times \sigma_{va} = \sigma_{va} \times C_3 \times \sigma_{va} = \sigma_{va} \times \sigma_{vc} = C_3^2$$
$$\sigma_{vb}^{-1} \times C_3 \times \sigma_{vb} = \sigma_{vb} \times \sigma_{va} = C_3^2$$
$$\sigma_{vc}^{-1} \times C_3 \times \sigma_{vc} = \sigma_{vc} \times \sigma_{vb} = C_3^2$$

4. FLOW CHART FOR DETERMINATION OF MOLECULAR POINT GROUPS

Using a flow chart, shown in Figure 25, it is easy to determine the point groups of different molecules, provided the geometry of the molecule is known to the applicant.

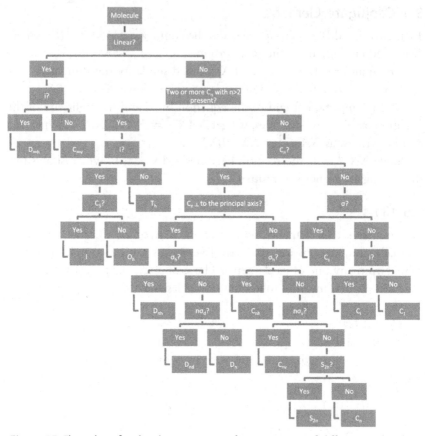

Figure 25 Flow chart for the determination of point groups of different molecules.

5. EXERCISES

5.1 Multiple Choice Questions

1. Which of the following is the principal axis of rotation in a NH_3 molecule?

   (a) C_1 (b) C_2
   (c) C_3 (d) C_4

2. Which of the following are the principal axes of rotation in H_2O, CH_4, SF_6, cyclohexane and (O_2, N_2, HCl, CO, NO) molecules, respectively?

   (a) C_2, C_3, C_4, C_6, C_1 (b) C_2, C_3, C_4, C_2, C_∞
   (c) C_2, C_3, C_4, C_6, C_∞ (d) C_2, C_3, C_5, C_6, C_∞

3. Which planes of symmetry are present in H_2O, NH_3, allene, *trans*-dichloroethylene and (HCl, CO, NO, HCN) molecules, respectively?

(a) $\sigma_d, \sigma_v, \sigma_d, \sigma_h, \sigma_v$ (b) $\sigma_v, \sigma_h, \sigma_d, \sigma_h, \sigma_v$

(c) $\sigma_v, \sigma_v, \sigma_d, \sigma_h, \sigma_v$ (d) $\sigma_v, \sigma_v, \sigma_d, \sigma_h, \sigma_d$

4. For which of the values of n, $S_n^n = E$?

(a) odd   (b) odd and even

(c) even   (d) none

5. If n is an odd number, what will be the value of S_n^n ?

(a) E (b) C_n

(c) σ_h (d) σ_v

6. $S_n^n = ?$

(a) i (b) E

(c) $C_n^n \times \sigma_h^n$ (d) none

7. Which of the following geometries contain the V_∞ axis of symmetry?

(a) tetrahedral   (b) square pyramidal

(c) linear   (d) octahedral

5.2 Short/Long Answer Questions

1. Identify the symmetry elements and detect the point groups of the following molecules and ions: ICl_4^-, C_2H_4, C_4H_8 (Cyclobutane), C_5H_5N (Pyridine), CO_2, CCl_4, 1:3:5 trichlorobenzene, CH_4, $Fe(CN_6)^{-4}$, $B_6H_{12}^{-2}$, napthalene, HCN, HCl, CO_2, O_2, N_2, NH_3.

2. Define the following terms: symmetry operation, symmetry element, principal axis, identity operation, improper rotation, inversion, symmetry group, point group, conjugate elements, similarity transformation and class.

3. Explain the following by giving suitable illustrations: rotation, reflection, inversion and improper rotation.

4. List the rules that the symmetry elements of a point group must obey.

5. Give an account of C_n, C_{nv}, D_n, cubic point groups and D_{nh} groups.

6. 'A group of six elements can have subgroups of order 1 or 2 or 3 only': Explain.

SUGGESTED FURTHER READINGS

The topics discussed in this chapter are a part of standard graduate curriculum. The majority of the textbooks with titles related to organic chemistry, inorganic chemistry and theoretical chemistry can act as a source of further reading. Moreover, there are several web resources useful for further learning. Some of them are listed below.

https://en.wikipedia.org/wiki/Symmetry
http://www.adrianbruce.com/Symmetry/
http://chemistry.rutgers.edu/undergrad/chem207/SymmetryGroupTheory.html
http://chemwiki.ucdavis.edu/Theoretical_Chemistry/Symmetry/Group_Theory%3A_Theory
http://vlab.amrita.edu/?sub=2&brch=193&sim=1013&cnt=1
http://nptel.ac.in/syllabus/104106044/
http://nptel.ac.in/courses/104106063/Module%201/Lectures%201-3/Lectures%201-3.pdf

CHAPTER 3

Isomerism in Coordination Complexes

Contents

Essentials of Coordination Chemistry
http://dx.doi.org/10.1016/B978-0-12-803895-6.00003-3

1. INTRODUCTION

There are several different entities in the world that look alike. The inability to distinguish these things is due to a lack of observation. For example, a layman hardly sees the difference between two sheep or cows, per se. However, the owner of these animals observes them carefully and distinguishes each of them from a crowd. Superimposing two similar entities and attempting to distinguish them is one of the ways of knowing whether the two moieties are the same or not. However, to be sure, one needs to substitute one of the entities for the other and inspect its functionality in every aspect. Though both the left and right hand or leg look similar, it is impossible to replace the left hand or leg with the right hand or leg of a human being and obtain the same functionality, or vice versa. In the same way, the right-leg shoe, right-hand glove or a right-side mirror of a vehicle do not fit appropriately on the counter side. The molecular world is full of such entities.

Two or more different compounds can have the same molecular formula. Hence, molecular formula alone is not sufficient to define a particular coordination compound. These different compounds with the same molecular formula are called isomers, and the phenomenon is called isomerism.

In 1823, J. von Liebig first showed that the silver salt of fulminic acid, $Ag-O-N\equiv C$, and silver isocyanate, $Ag-N\equiv C\equiv O$, have the same composition but different properties. However, the name 'isomerism' was first given by J. Berzelius in 1830. The phenomenon of isomerism was successfully explained first by the theory of chemical structure developed by A.M. Butlerov in the 1860s.

Due to a wide variety of geometries and coordination numbers, the metal complexes show a variety of isomerism. Sometimes, these isomers undergo a rapid interconversion. The isomers can only be identified if the speed of the instrumental technique used is faster than that of the interconversions. Thus, with the inventions of sophisticated analytical instruments, there is a possibility of the discovery of new isomers. The isomers may or may not show the same physical and chemical properties. The application of molecules for particular purposes often requires typical structural arrangement within a molecule. Thus, a study of the structure activity relationship, abbreviated as SAR, is becoming popular. The complex $[Pt(Cl)_2(NH_3)_2]$, commonly known as platin, shows two isomers; one of the isomers, known as *cis*-platin, is found to be very effective against tumours, whereas the *trans*-platin is not useful for the treatment of tumours. The studies have shown that the *cis*-platin binds to the strands of the malicious DNA and breaks its quaternary or tertiary structure (denaturation), leading to the prevention of further growth. It has also been proven that if $\angle Cl-M-Cl$ is below $95°$, the required binding with DNA is permitted. The $\angle Cl-M-Cl$ in *cis* platin is found to be $92°$, whereas in the case of *trans*-platin, it is around $180°$. Thus, the knowledge of isomerism and the capability of synthesizing desired isomers is important in applying the molecules to different purposes. In the present chapter, a wide variety of isomerism in coordination compounds has been included. The categories discussed here are not mutually exclusive, and hence, two or more types of isomerism may have to be invoked to fully describe the isomerism between complexes. Methods of structure determination, such as X-ray diffraction, have been useful in identifying many isomers of coordination compounds. Typical reactions for distinguishing geometrical isomers are also discussed.

2. CLASSIFICATION OF ISOMERISM

In order to provide a detailed discussion on isomerism among metal complexes, the isomers are first broadly classified on the basis of topology as structural isomerism and stereoisomerism, as shown in Figure 1. The stereoisomers have the same topology; it is the arrangements of the ligands in the space that make them different. The structural isomerism, also known as constitutional isomerism, includes all other types of isomerism shown by the metal complexes. The structural isomers have different topology.

A detailed classification of isomerism in metal complexes is given in Figure 2. The structural or constitutional isomerism is further divided into at least 10 subclasses. Some of these classes are not in strict adherence with the definition given above. The further classification of stereoisomerism

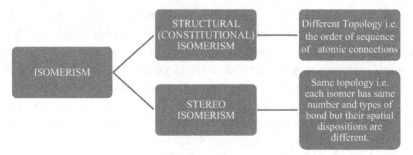

Figure 1 Broad classification of isomerism in coordination complexes.

gives rise to geometrical isomerism and optical isomerism. Individual types of isomerism with suitable examples are discussed in the following sections of the chapter.

3. STRUCTURAL ISOMERISM

3.1 Conformation Isomerism

The structural isomers that differ only with respect to a stereochemistry point of view are named as conformation isomers.

Complexes of only a few metal ions show this type of isomerism. The d^8 metal ions exhibit conformation isomerism. A diphenylphosphino methane (dppm) complex $[NiCl_2(Ph_2PCH_2PPh_2)_2]$ existing in both *trans*-square planar and tetrahedral stereochemistry, as shown in Figure 3, is a typical example of conformation isomerism.

Sometimes, the difference in stereochemistry results in a difference in the magnetic properties of a complex. Square planar complexes of Ni(II) are diamagnetic, whereas the tetrahedral ones are paramagnetic. Hence, some of the conformation isomers are also spin isomers.

3.2 Spin Isomerism

Structural isomerism observed due to different electron spins in complexes of the same geometry is known as spin isomerism. In some of the complexes of Fe(II), Fe(III) and Co(II), the metal ions are known to exist in both high-spin and low-spin states. Fe(III) ion in an octahedral ligand field can have either a high-spin $(t_{2g})^5 (e_g)^0$ or low-spin $(t_{2g})^3 (e_g)^2$ configuration, as shown in Figure 4.

Spin isomerism is known to be present in a complex $[Fe^{+3}(S_2CNMe_2)_3]$, as shown in Figure 5.

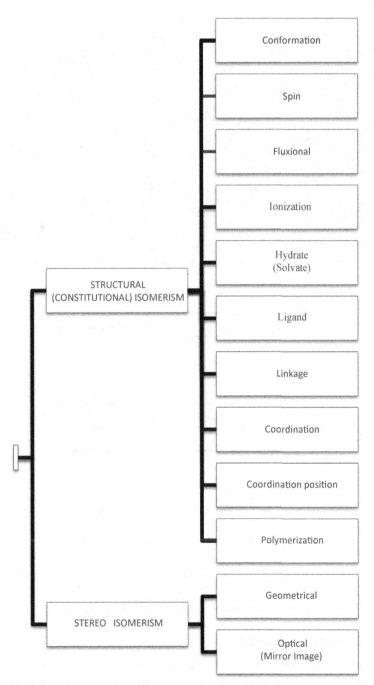

Figure 2 Detailed classification of isomerism in coordination complexes.

(a)   (b)

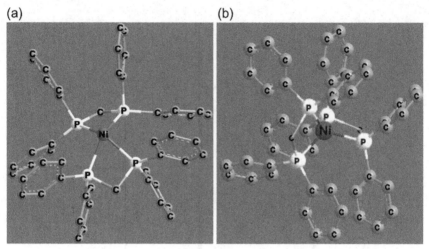

Figure 3 (a and b) Conformation isomers of Ni(II)(dppm) complex.

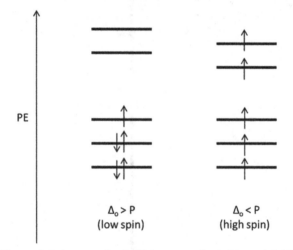

Figure 4 Spin states for Fe(III) in an octahedral ligand field.

Spin isomers coexist in the sample, but their lifetimes are very short ($\sim 10^{-7}$ s) and hence are very difficult to distinguish. The spin isomerism behaviour in solution and solid states are found to be different.

3.3 Fluxional Isomerism

Fluxional molecules are the molecules that rapidly undergo intramolecular rearrangements among their constituent atoms. Such phenomenon exhibited by ketones/phenols is termed as tautomerism in organic chemistry.

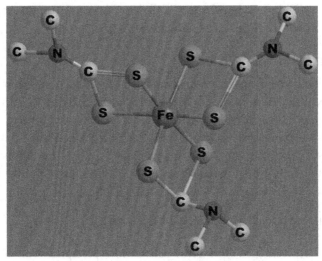

Figure 5 Structure of [Fe(S₂CNMe₂)₃].

The coexistance of dicobaltoctacarbonyl [Co$_2$(CO)$_8$] in two isomeric structures is depicted in Figure 17 of Chapter 8. The molecule η^5-C$_5$H$_5$Mn(CO)$_3$, shown in Figure 6, is capable of showing 15 fluxional isomers.

From the structure, it is seen that the unit η^5-C$_5$H$_5$Mn has a five-fold rotational axis due to the cyclopentadienyl ring. Similarly, the unit Mn(CO)$_3$ possess a three-fold rotational axis due to the three carbonyl groups. Combining these two sets of axes, there are, in all, 15 rotational isomers (fluxional isomers) possible in this molecule. A 15-fold rotational axis affords an angle of 24° (360°/15 = 24°) between two equipotential arrangements. A negligible energy change is involved during the interchange.

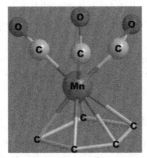

Figure 6 Structure of η^5-C$_5$H$_5$Mn(CO)$_3$.

Thus, the individual nomenclatures of all these structural isomers are avoided and are collectively named as fluxional molecules.

3.4 Ionization Isomerism

Isomers that give different ions in the solution are called ionization isomers.

An example of a pair of ionization isomers is $[Co(NH_3)_5Br]SO_4$ & $[Co(NH_3)_5SO_4]Br$. These two isomers can be easily identified by performing tests for sulphate and bromide ions in their aqueous solutions.

$[Co(NH_3)_5Br]SO_4$ gives white precipitates of barium sulphate with an aqueous solution of barium chloride. It does not give the test for bromide ion.

$$[Co(NH_3)_5Br]SO_4 \rightarrow [Co(NH_3)_5Br]^{+2}_{(aq)} + SO_4^{-2}{}_{(aq)}$$

While $[Co(NH_3)_5SO_4]Br$ gives pale yellow precipitates of silver bromide with an aqueous solution of silver nitrate, it does not give the test for sulphate ion.

$$[Co(NH_3)_5SO_4]Br \rightarrow [Co(NH_3)_5SO_4]^{+1}_{(aq)} + Br^-{}_{(aq)}$$

A few more examples of ionization isomers are:

$$[PtBr(NH_3)_3]NO_2 \text{ and } [Pt(NO_2)(NH_3)_3]Br$$

$$[Pt(NH_3)_4Cl_2]Br_2 \text{ and } [Pt(NH_3)_4 Br_2]Cl_2$$

$$[Co(en)_2NO_2Cl]SCN \text{ and } [Co(en)_2NO_2SCN]Cl$$

The structural formulae of ionization isomers can be deduced by the measurement of molar conductance. Consider the following example:

If a complex with the molecular formula $CrCl_3(H_2O)_4$ exhibits molar conductance corresponding to two ions, the dissociation can only be written as $[Cr(H_2O)_4Cl_2]Cl \rightarrow [Cr(H_2O)_4Cl_2]^+ + Cl^-$.

The above formula can be verified by measuring the amount of precipitation of AgCl using $AgNO_3$.

A peculiar variation in this type of isomerism is shown by $[Cu(NH_3)_5(SO_3)]NO_3$ and $[Co(NH_3)_5NO_2]SO_4$.

3.5 Hydrate (Solvate) Isomerism

The general name for this isomerism is solvate isomerism. Specifically in aqueous solutions, the isomerism is referred to as hydrate isomerism.

The isomers in which coordinated ligands are replaced by water (solvent) molecules are known as hydrate (solvate) isomers.

Green crystals of chromium(III) chloride are formed in a hot aqueous solution upon reduction of chromium(IV) oxide with hydrochloric acid. When these crystals are dissolved in water, the chloride ligands in the complex are slowly replaced by water to give blue-green and finally violet-coloured complexes.

One g quantity of each of these three dark green, blue-green and violet-coloured isomers of complexes with the molecular formula $CrCl_36H_2O$, when treated with dehydrating agents, attain a constant weight of 0.865, 0.932 and 1 g, respectively.

Using this information, it is possible to identify the structural formula of the isomers as follows:

The molecular weight of $CrCl_36H_2O = 52 + (35.5 \times 3) + (6 \times 18) = 266.5$ g

1 g of dark green isomer loses $(1 - 0.865) = 0.135$ g of H_2O

Therefore, 266.5 g of dark green isomer loses $(0.135 \times 266.5) = 36.0$ g of H_2O, which corresponds to two water molecules. Thus, the dark green isomer has two water molecules outside the coordination sphere of the complex.

Similarly,

1 g of blue-green isomer loses $(1 - 0.932) = 0.068$ g of H_2O

Therefore, 266.5 g of dark green isomer loses $(0.068 \times 266.5) = 18.12$ g of H_2O, which corresponds to one water molecule. Thus, the blue-green isomer has only one water molecule outside the coordination sphere of the complex.

Since, the violet isomer does not lose any weight, there is no water molecule in the outer shell of this isomer.

Also, when 1 g of these three dark green, blue-green and violet-coloured isomers of a complex with the molecular formula $CrCl_36H_2O$ are treated with an excess of silver nitrate, the samples precipitate AgCl in the amount of 0.538, 1.076 and 1.614 g, respectively.

Using this information, it is possible to identify the structural formula of the isomers as follows:

The molecular weight of $CrCl_36H_2O = 52 + (35.5 \times 3) + (6 \times 18) = 266.5$ g

The molecular weight of $AgCl = 108 + 35.5 = 143.5$

1 g of dark green isomer gives 0.538 g AgCl.

Table 1 Hydrate isomers of aquated chromium(III) chloride

$[Cr(H_2O)_6]Cl_3$	$[Cr(H_2O)_5Cl]Cl_2H_2O$	$[Cr(H_2O)_4Cl_2]Cl_2H_2O$
Violet	Blue green	Dark green
• Gives three chloride ions in aqueous solution • No loss of moisture upon dehydration	• Gives two chloride ions in aqueous solution • Loss of moisture corresponding to two water molecules upon dehydration	• Gives one chloride ion in aqueous solution • Loss of moisture corresponding to three water molecules upon dehydration

Therefore, 266.5 g of dark green isomer gives $(0.538 \times 266.5) =$ 143.37 g of AgCl, which corresponds to one Cl^- ion. Thus, the dark green isomer has one chloride ion outside the coordination sphere of the complex.

Similarly,

1 g of blue-green isomer gives 1.076 g AgCl.

Therefore, 266.5 g of dark green isomer loses $(1.076 \times 266.5) = 86.7$ g of AgCl, which corresponds to two Cl^- ions. Thus, the blue-green isomer has two chloride ions outside the coordination sphere of the complex.

And,

1 g of violet isomer gives 1.614 g AgCl.

Therefore, 266.5 g of dark green isomer loses $(1.614 \times 266.5) =$ 430.13 g of AgCl, which corresponds to three Cl^- ions. Thus, the violet isomer has three chloride ions outside the coordination sphere of the complex.

From the above calculations, the structural formula of the three isomers are obtained and shown in Table 1:

Other examples of hydrate isomerism are the following:

$$[CoCl(H_2O)en_2Cl_2 \quad \text{and} \quad [CoCl_2en_2]ClH_2O,$$

$$[CrCl_2(H_2O)_2Py_2]Cl \quad \text{and} \quad [CrCl_3(H_2O)_2Py_2]H_2O.$$

3.6 Ligand Isomerism

The complexes of isomeric ligands are known as ligand isomers.

1,2-diaminopropane and 1,3-diaminopropane form complexes showing ligand isomerism, as shown in Figure 7.

(a)   (b)

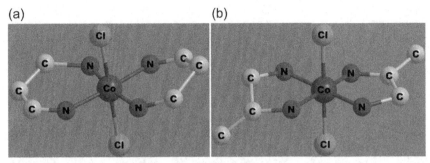

Figure 7 (a) 1,3-diaminopropane and (b) 1,2-diaminopropane ligand isomer complexes.

Other complexes showing ligand isomerism [Cu(3,4-lutidine)$_4$Cl$_2$] and [Cu(3,5-lutidine)$_4$Cl$_2$], have been isolated using thin layer chromatography as bright blue and pale blue coloured complexes, respectively.

In the case of certain mixed ligand complexes, the total elemental composition due to the co-ligands becomes the same. These type of ligand isomers are known as co-ligand isomers.

Generally, a pair of co-ligand isomers results due to an intramolecular nucleophilic addition reaction in coordinated ligands. One of the isomers is an ordinary complex, while the other isomer formed upon the above-mentioned reaction is a chelate.

An illustration of co-ligand isomerism [1] in octahedral complexes is provided by *cis*-[Co(en)$_2$(NH$_2$CH$_2$CN)Cl]$^{+2}$ which converts into its chelate form of the co-ligand isomer; [Co(en)(NH$_2$CH$_2$C(NH$_2$)=NCH$_2$CH$_2$NH$_2$) Cl]$^{+2}$ in nearly neutral or basic solution, as shown in Figure 8.

(a)   (b)

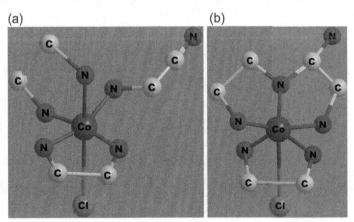

Figure 8 (a) *Cis*-[Co(en)$_2$(NH$_2$CH$_2$CN)Cl]$^{+2}$ and (b) [Co(en)(NH$_2$CH$_2$C(NH$_2$)=NCH$_2$CH$_2$NH$_2$)Cl]$^{+2}$.

(a)   (b)

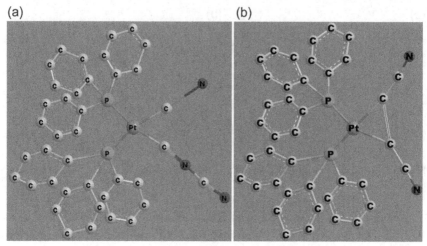

Figure 9 (a) [Pt(PPh$_3$)$_2$(TCNE)] and (b) co-ligand isomer.

Yet another example of co-ligand isomerism [1] in square planar complexes where irradiation leads to the formation of a complex from a chelate is provided by [Pt(PPh$_3$)$_2$(TCNE)], as shown in Figure 9.

3.7 Linkage Isomerism

The ligands that are capable of binding the metal ion using either of the more than one coordinating atoms are called ambidentate ligands. NO_2^- is an ambidentate ligand, as it can coordinate either through a nitrogen or oxygen atom. SCN^- and $S_2O_3^{-2}$ are also ambidentate ligands.

The isomers formed when ambidentate ligands use different coordinating atoms for binding the metal ion are known as linkage isomers. Such a pair of isomeric complexes with a NO_2^- ligand is shown in Figure 10.

(a)   (b)

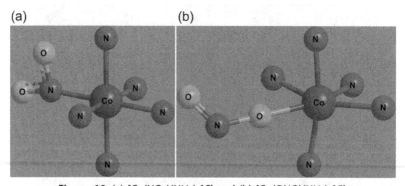

Figure 10 (a) [Co(NO$_2$)(NH$_3$)$_5$]Cl and (b) [Co(ONO)(NH$_3$)$_5$]Cl.

(a)   (b)

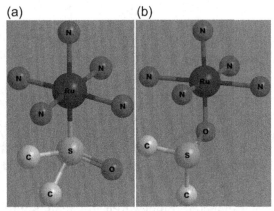

Figure 11 (a) S- and (b) O-bonded linkage isomers of $[Ru(NH_3)_5(dmso)]^{3+}$.

Octahedral Ruthenium(II) and (III) complexes $[Ru(NH_3)_5(dmso)]^{2+/3+}$ exhibit linkage isomerism [2] as both sulphur and oxygen are potential coordinating atoms in dimethyl sulfoxide, as shown in Figure 11.

3.8 Coordination Isomerism

Structural isomers having both cationic and anionic parts as complexes with a different distribution of ligands and/or metal ions are known as coordination isomers.

The pair $[Cr(NH_3)_6]^{+3}[Cr(SCN)_6]^{-3}-[Cr(NH_3)_4(SCN)_2]^{+}[Cr(NH_3)_2(SCN)_4]^{-}$ involves the same metal ion (Cr^{+3}) in both the cationic and anionic complex ions.

While the pair $[Co(NH_3)_6]^{+3}[Cr(CN)_6]^{-3}-[Cr(NH_3)_6]^{+3}[Co(CN)_6]^{-3}$ illustrates coordination isomerism involving complexes with different metal ions.

A typical example of coordination isomerism shown by complex ions with same metal ions in different oxidation states is provided by $[Pt(py)_4]^{+2}[Pt(Cl)_6]^{-2}-[Pt(py)_4Cl_2]^{+2}[PtCl_4]^{-2}$.

A pair of square planar complexes, $[Cu(NH_3)_4]^{+2}[Pt(Cl)_4]^{-2}-[Pt(NH_3)_4]^{+2}[Cu(Cl)_4]^{-2}$, is an example of coordination isomerism showed by a square planar four-coordinate complex ion pair.

3.9 Coordination Position Isomerism

Isomers of bridged polynuclear complexes with a different distribution of non-bridging ligands with respect to the metal ion are known as coordination position isomers. Symmetrical and asymmetrical distribution of ligands in a dinuclear complex is shown in Figure 12.

(a)   (b)

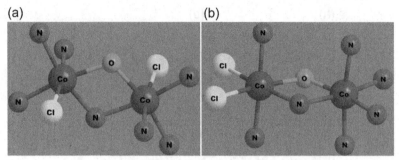

Figure 12 (a) Symmetrically and (b) asymmetrically bonded non-bridging ligands.

A heterometallic dinuclear complex $[(Ph_3P)_2ClPd(CS_2)Pt(PPh_3)_2]$ $[BF_4]$ also shows coordination position isomerism [3] as shown in Figure 13.

Polymerization isomerism represents the variable value of 'x' in a complex $[ML_n]_x$. Actually, it is not an isomerism, but it is included in this list because it represents an additional way in which an empirical formula may give incomplete information about the nature of a complex. All members of the series; $[Co(NH_3)_3(NO_2)_3]_x$ are polymerization isomers, as shown in Table 2.

4. STEREO ISOMERISM OR SPACE ISOMERISM

The isomers that differ with respect to the relative orientation of bonds are known as stereo isomers or space isomers. There are two forms of stereo-isomerism: geometrical isomerism, also known as diastereoisomerism, and Optical isomerism, also known as mirror image isomerism.

4.1 Geometrical Isomerism: Four-Coordinate Complexes

There are two possible geometries for complexes of coordination number four. These geometries are two-dimensional square planar and three-dimensional tetrahedral. In the square planar complexes, the $\angle L-M-L$ can be $90°$ or $180°$. The angle involving adjacent ligands is $90°$, while the angle involving oppositely placed ligands is $180°$. The $\angle L-M-L$ involving any two ligands in the case of tetrahedral complexes are $109.5°$, which makes the entire four positions equivalent. The geometrical isomers are essentially distinguished using the above-mentioned angles.

4.1.1 Tetrahedral Complexes

Since all the four positions in a tetrahedron are equivalent, tetrahedral complexes containing monodentate ligands a, b, c and d of all the types, viz,

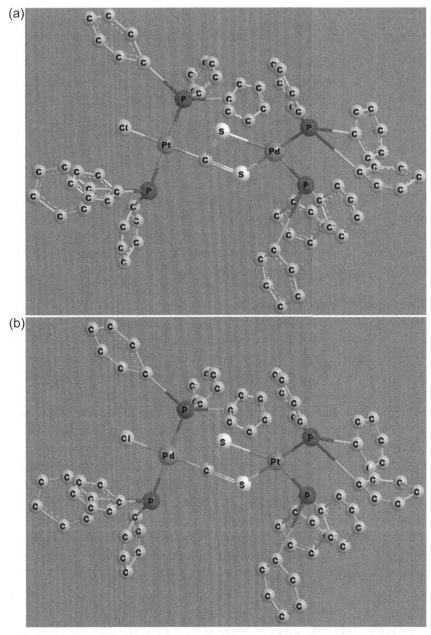

Figure 13 (a and b) Coordination position isomers of [(Ph₃P)₂ClPd(CS₂)Pt(PPh₃)₂].

Table 2 Polymerization isomers of $[Co(NH_3)_3(NO_2)_3]_x$

Sr. No.	Empirical formula	Value of 'x'
1.	$[Co(NH_3)_3(NO_2)_3]$	1
2.	$[Co(NH_3)_6]\ [Co(NO_2)_6]$	2
3.	$[Co(NH_3)_4(NO_2)_2]\ [Co(NH_3)_2(NO_2)_4]$	2
4.	$[Co(NH_3)_5(NO_2)]\ [Co(NH_3)_2(NO_2)_4]_2$	3
5.	$[Co(NH_3)_6]\ [Co(NH_3)_2(NO_2)_4]_3$	4
6.	$[Co(NH_3)_4(NO_2)_2]_3\ [Co(NO_2)_6]$	4
7.	$[Co(NH_3)_5(NO_2)_6]_3\ [Co(NO_2)_6]_2$	5

[Ma$_4$], [Ma$_2$b$_2$], [Ma$_3$b] and [Mabcd] can exist in only one geometrical form. They do not exhibit geometrical isomerism. Even polydentate ligands alone or in combination with monodentate ligands cannot alter the relative orientations of the bonds in tetrahedral complexes. Hence, tetrahedral complexes do not show geometrical isomerism.

4.1.2 Square Planar Complexes
The two distinguished forms of $[Pt(NH_3)_2Cl_2]$ led a surge in proposing a square planar structure for four coordinated complexes of Pt(II) ion. A systematic investigation revealed that square planar complexes of type [Ma$_2$b$_2$] can exist in 1,2 (*cis-*) and 1,3 (*trans-*) form. The terms *cis-* and *trans-* refer to the positions of two groups with respect to each other. In the *cis-*isomer, they are next to each other or at an angle of 90° with respect to the central metal ion, whereas in the *trans-*isomer they are opposite each other, i.e. at 180° relative to the central metal ion, as shown in Figure 14.

Tsherniaev isolated the three forms predicted for the planar species [Mabcd] of the complex $[Pt(NH_3)(NH_2OH)Py(NO_2)]Cl$, which added interest to Werner's postulation. The planar structure of four-coordinate complexes has since been confirmed by physical methods of structure determination such as dipole moment measurement, vibrational spectroscopy and X-ray diffraction. Complexes with square planar geometry are usually formed by the central metal ions such as Co(II), Ni(II), Cu(II),

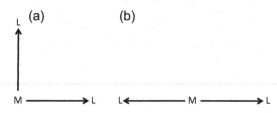

Figure 14 (a) *Cis-* and (b) *trans-*positions in square planar geometry.

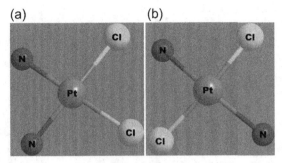

Figure 15 (a) *Cis*- and (b) *trans*-isomers of [Pt(NH$_3$)$_2$Cl$_2$].

Au(III) and Pd(II) with d^7, d^8 or d^9 configuration. The complexes of type [Ma$_4$] or [Ma$_3$b] do not render any stereoisomers, as all the possible arrangements of a and b for each of these types are exactly equivalent.

A detailed treatise on stereoisomerism in different types of square planar complexes is provided in the following section.

4.1.2.1 [Ma$_2$b$_2$]$^{\pm n}$ Type

Here M is a central ion, while a and b are monodentate ligands. Complexes of this type exist as *cis–trans*-isomers. As discussed in the introductory section, the *cis*-isomer of the complex [Pt(NH$_3$)$_2$Cl$_2$], also known as *cis*-platin, is a popular antitumour drug.

As shown in Figure 15, in such isomers, all the four bonds are coplanar with the central atom.

Similarly, [Pd(NH$_3$)$_2$(NO$_3$)$_2$] also shows *cis*- and *trans*-isomers.

4.1.2.2 [Ma$_2$bc]$^{\pm n}$ Type

In this type of complex, 'a' is any neutral monodentate ligand, such as NH$_3$, py, H$_2$O, while 'b' and 'c' are anionic monodentate ligands, such as Cl$^-$, Br$^-$ and NO$_2{}^-$. The isomers are distinguished on the relative positions of the ligand 'a', as illustrated by [Pt(NH$_3$)$_2$ClNO$_2$] in Figure 16.

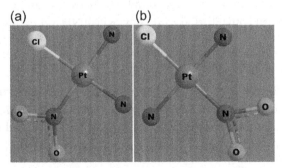

Figure 16 (a) *Cis*- and (b) *trans*-isomers of [Pt(NH$_3$)$_2$ClNO$_2$].

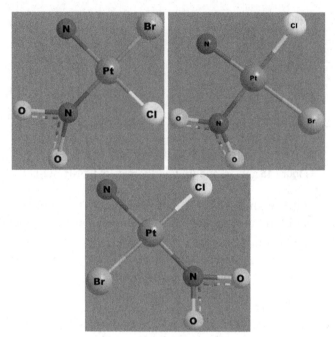

Figure 17 Three isomeric forms of $[Pt(NH_3)ClBrNO_2]$.

4.1.2.3 $[Mabcd]^{\pm n}$ Type

The Pt(II) ion forms a number of complexes of this type. Complexes of this type exist in three isomeric forms. These isomers are conveniently derived by fixing the position of one of the ligands, say 'a', and then putting the remaining ligands, viz, 'b', 'c' and 'd', one by one at any other fixed position. In this manner, three distinct geometrical arrangements can be obtained. Any other isomer obtained using a trial and error method will certainly be indistinguishable from at least one of the three isomers, shown in Figure 17.

A few more examples of complexes showing three geometrical isomers are $[Pt(NH_3)pyClBr]$, $[Pt(NO_2)py(NH_3)(NH_2OH)]^{+}$ and $[Pt(C_2H_4)(NH_3)ClBr]$.

4.1.2.4 $[M(AB)_2]^{\pm n}$

Polydentate ligands are capable of introducing more variations in isomerism. Complexes containing unsymmetrical bidentate chelating ligands can also exist as *cis-* and *trans-*isomers of type $[M(AB)_2]^{\pm n}$, where 'AB' is unsymmetrical bidentate chelating ligand characterized by different terminal atoms 'A' and 'B'. As the amino acids possess two different terminal groups, they will serve as the best ligands to illustrate isomerism in this type of complex. [Pt(gly)$_2$], where gly is $NH_2-CH_2COO^-$ (glycine ion), exists in the following *cis–trans*-isomeric forms, as shown in Figure 18.

Yet another example is provided by square planar Ni(II) thioselenophosphinate, Ni(SeSPPh$_2$)$_2$ complex [4], as shown in Figure 19.

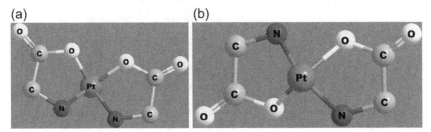

Figure 18 (a) *Cis*- and (b) *trans*-isomers of [Pt(gly)$_2$].

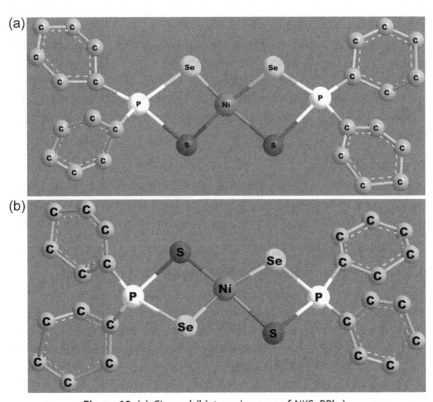

Figure 19 (a) *Cis*- and (b) *trans*-isomers of Ni(SePPh$_2$)$_2$.

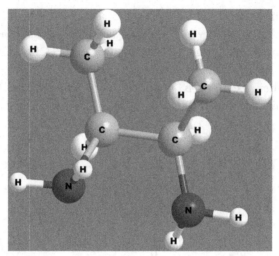

Figure 20 Symmetrical bidentate ligand: $NH_2CH(CH_3)CH(CH_3)NH_2$.

4.1.2.5 $[M(AA)_2]^{\pm n}$

Complexes containing symmetrical bidentate chelating ligands can also exist as *cis-* and *trans-*isomers. The complex $[M(AA)_2]^{\pm n}$ represents a square planar complex with 'AA' as a symmetrical bidentate chelating ligand, shown in Figure 20. The complex ion with the formula $[Pt(NH_2CH(CH_3)CH(CH_3)NH_2)_2]^{+2}$ illustrates geometrical isomerism in such complexes, as shown in Figure 21.

4.1.2.6 $M_2a_2b_4$ Type

A few more complexities in the geometrical isomerism are observed in binuclear complexes. In these types of complexes, *cis-* and *trans-*isomers, as well as the asymmetrical form, also exist, as shown by the complex $[Pt(PEt_3)Cl_2]_2$ illustrated in Figures 22 and 23.

4.2 Geometrical Isomerism: Six-Coordinate Complexes

The majority of the metals form at least one six-coordinate complex. The coordination number six, being the most common, is the most extensively studied coordination number in coordination chemistry. Three possible arrangements of six-monodentate ligands around the central metal ion are possible. These three arrangements lead to geometries such as hexagonal planar, trigonal prismatic and octahedral, as shown in Figure 24.

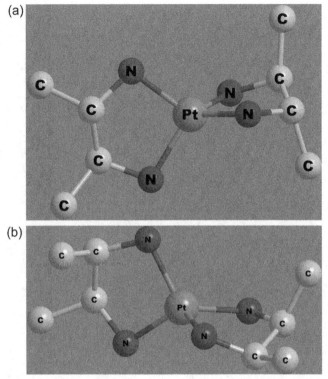

Figure 21 (a) *Cis-* and (b) *trans-*isomers of $[Pt(NH_2CH(CH_3)CH(CH_3)NH_2)_2]^{+2}$.

The physical and chemical evidence, especially those based on isomerism, have shown that the most favoured arrangement of six ligands in a complex is always octahedral with the entire six positions equivalent. The arrangement of six ligands in a regular octahedral complex around the central metal ion can be represented as shown in Figure 25.

The complexes with the general formula $[Ma_6]^{\pm n}$ and $[Ma_5b]^{\pm n}$ do not show geometrical isomerism. The following types of complexes show geometrical isomerism.

4.2.1 $[Ma_4b_2]^{\pm n}$ Type

Geometrical isomerism in this type of complex is easily identified by focussing on the relative positions of the minority ligand 'b'. If the two 'b' ligands occupy adjacent positions ($\angle b-M-b = 90°$), the isomer is named as a *cis*-isomer. In the case of a *trans*-form, the two 'b' ligands are diagonally opposite to each other ($\angle b-M-b = 180°$), as shown in Figure 26.

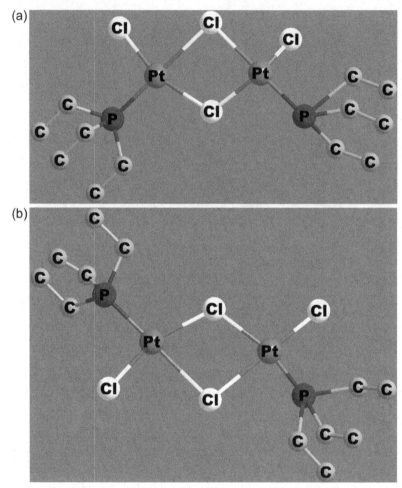

Figure 22 (a) *Cis-* and (b) *trans*-isomers of [Pt(PEt$_3$)Cl$_2$]$_2$.

In other words, in *cis*-form 'b', ligands lie on any of the 12 edges of the octahedron, and in *trans*-form, 'b' ligands are at the end of the straight line passing through the central atom. [Co(NH$_3$)$_4$Br$_2$]$^+$ is a well-known example for this type. It is interesting to note here that out of a total 15 possible arrangements, 12 arrangements render a *cis*-isomer, while the remaining three arrangements give *trans*-isomers. In *cis*-isomers, two Cl$^-$ ions are at the adjacent position (i.e. 1,2 / 1,3 / 1,4 / 1,5 / 2,3 / 2,5 / 2,6 / 3,4 / 3,5 / 3,6 / 4,5 / 4,6 –positions), while in *trans*-isomers, two Cl$^-$ ions are opposite to each other (i.e. 1,6 / 2,4 / 3,5 –positions), as shown in Figure 26.

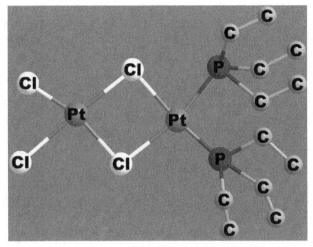

Figure 23 Asymmetrical form of [Pt(PEt$_3$)Cl$_2$]$_2$.

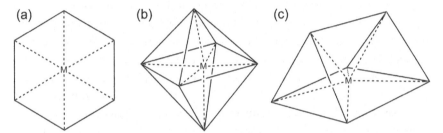

Figure 24 (a) Hexagonal planar (b) octahedral (c) trigonal prismatic geometries.

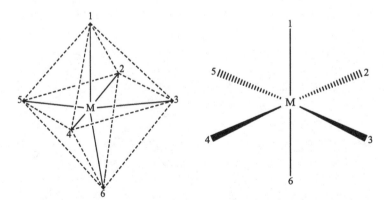

Figure 25 Arrangement of six ligands in an octahedral complex.

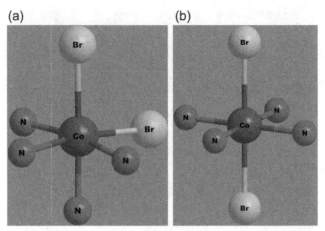

Figure 26 (a) *Cis*- and (b) *trans*-isomers of [Co(NH₃)₄Br₂].

4.2.2 [Ma₃b₃]±ⁿ Type

In order to classify the various arrangements into *cis*- and *trans*-isomers in this type of complex, consider the following. The total number of possible arrangements of ligands will be the same as the total number of ways in which the three ligands 'a' can be arranged in an octahedron. The number of possible arrangements can be given mathematically using the principle of combination ^nC_r, where 'n' is the total number of available positions (six in the case of an octahedron), while 'r' is the number of 'a' ligands to be placed in the octahedron. Also, $^nC_r = n!/r! \times (n - r)!$ becomes $^6C_3 = 6!/3! \times (6 - 3)! = 20$. Out of these 20 arrangements, 12 arrangements will have at least one $\angle a{-}M{-}a = 180°$. All these arrangements correspond to the *trans*-isomers, while the remaining eight arrangements with all the three $\angle a{-}M{-}a$ as 90° are known as *cis*-isomers. Two such isomers are shown in Figure 27.

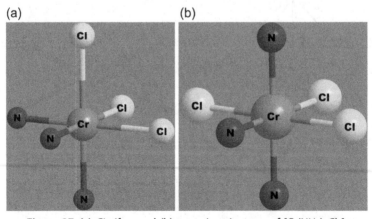

Figure 27 (a) *Cis*-/*fac*- and (b) *trans*-/*mer*-isomers of [Cr(NH₃)₃Cl₃].

In Figure 27, in the case of *cis*-isomers, three 'a' occupy 1, 2 and 3 positions, while in *trans*-isomers, they occupy 1, 2 and 6 positions. Such isomerism is illustrated by complexes $[Cr(NH_3)_3Cl_3]$ and $[Rh(py)_3Cl_3]$.

In a *cis*-isomer, three Cl^- ions are on the triangular face of the octahedron and three NH_3 molecules are on the opposite triangular face of the octahedron, hence it is called a facial isomer, abbreviated as *fac-*. While in the case of a *trans*-isomer, three Cl^- ions are around the edge of the octahedron and three NH_3 molecules are at the opposite edge of the octahedron, hence this is called a peripheral or meridianal isomer, abbreviated as *mer-*.

4.2.3 [Mabcdef]$^{\pm n}$ Type

In this type of complex, 720 arrangements yield 15 geometrical isomers. Only one compound known is $[Pt(py)(NH_3)(NO_2)ClBrI]$. Some of these were isolated and characterized by Anna Gel'man and reported in 1956, but no attempt has been made to isolate all of the isomers.

4.2.4 [M(AA)$_2$a$_2$]$^{\pm n}$ Type

Here (AA) is a symmetrical bidentate chelating ligand with ends A and 'a' representing two identical monodentate ligands. In a *cis*-isomer, 'a' ends are *cis-* to each other and in *trans*-isomers, 'a' ends are *trans-* to each other. *Cis-* and *trans*-forms of $[Co(en)_2(Cl)_2]^+$ are shown in Figure 28.

Illustrations for such species are also provided by complexes such as $[Co(en)_2(NO_2)_2]^+$, $[Ir(C_2O_4)_2Cl_2]^{2-}$, $[Ru(C_2O_4)_2Cl_2]^{2-}$, $[Cr(C_2O_4)_2(H_2O)_2]^-$, etc.

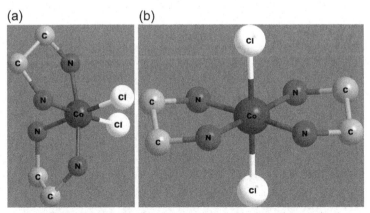

Figure 28 (a) *Cis-* and (b) *trans*-isomers of $[Co(en)_2(Cl)_2]^+$.

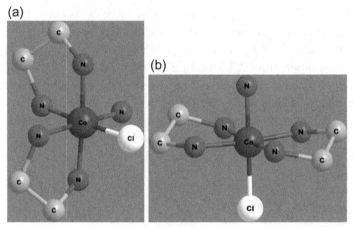

Figure 29 (a) *Cis*- and (b) *trans*-isomers of $[Co(en)_2(NH_3)Cl]^{2+}$.

4.2.5 $[M(AA)_2ab]^{\pm n}$ Type

Here (AA) is a symmetrical bidentate chelating ligand having ends A, while 'a' and 'b' are two different monodentate ligands. In a *cis*-isomer, 'AA' are *cis*- to each other, and in *trans*-isomers, 'AA' are *trans*- to each other. *Cis*- and *trans*-forms of $[Co(en)_2(NH_3)Cl]^{2+}$ are shown in Figure 29.

This type of isomerism is also shown by $[Ru(py)(C_2O_4)_2(NO)]$.

4.2.6 $[M(AA)a_2b_2]^{\pm n}$ Type

The ligands in this type of complex are as defined in the previous sections. The isomerism in such complexes is decided by the relative positions of the two 'a's and 'b's. $[Co(en)(NH_3)_2Cl_2]^{2+}$ illustrates the above mentioned isomerism, as shown in Figure 30.

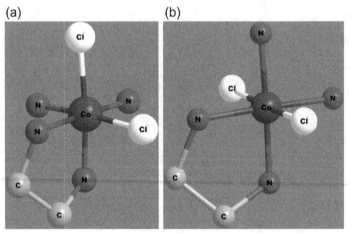

Figure 30 (a) *Cis*- and (b) *trans*-isomers of $[Co(en)(NH_3)_2Cl_2]^{2+}$.

(a)   (b)

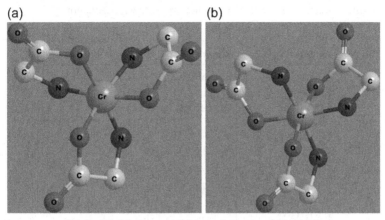

Figure 31 (a) *Cis-* and (b) *trans*-isomers of [Cr(gly)₃].

4.2.7 [M(AB)₃]^{±n} Type

Here, 'AB' is an asymmetrical bidentate chelating ligand. The geometrical isomerism in this type of complex is identified on the basis of relative positions of the atoms 'A' and 'B' with respect to each other. When the two 'A's or 'B's are placed opposite to each other, the isomer is *trans-*, while the other one is labelled as a *cis*-isomer. Such isomerism is illustrated by [Cr(gly)₃], as shown in Figure 31.

It is interesting to note here that the *cis*-isomer also refers to the *fac*-isomer corresponding to the arrangement of the 'A' or 'B' end of the asymmetrical bidentate ligand in a triangular position. Also, the *trans*-isomer is a *mer*-isomer in these cases.

4.3 Maximum Number of Geometrical Isomers in Different Complexes

Owing to various geometries and a wide variety of ligands, a detailed treatise on all types of complexes is avoided. However, the table shown below can provide a guide to studying isomerism and many other complexes. A student should prepare the hypothetical molecules and sketch the possible isomers for the complexes shown in Table 3.

5. EXPERIMENTAL SEPARATION AND IDENTIFICATION OF ISOMERS

5.1 Dipole Moment

The *cis-* and *trans*-isomers of Pt(II) complexes of the [PtA₂X₂] type (where A = substituted phosphine, arsine and X = halogen) have been

Table 3 Maximum number of geometrical isomers in different complexes

Sr. No.	Formula	Geometry	Maximum number of geometrical isomers
1.	Ma_4	Tetrahedral	1
	Mabcd		1
2.	Ma_4	Square planar	1
	Ma_3b		1
	Ma_2b_2		2
	Ma_2bc		2
	Mabcd		3
	$M(AA)_2$		2
	$M(AB)_2$		2
	$M_2A_2B_4$		3
3.	Ma_6	Octahedral	1
	Ma_5b		1
	Ma_4b_2		2
	Ma_3b_3		2
	Ma_4bc		2
	Ma_3bcd		5
	Ma_2bcde		15
	Mabcdef		30
	$Ma_2b_2c_2$		6
	Ma_2b_2cd		8
	Ma_3b_2c		3
	$M(AA)_2a_2$		2
	$M(AA)_2ab$		2
	$M(AA)a_2b_2$		2
	M(AA)(BC)de		10
	M(AB)(AB)cd		11
	M(AB)(CD)ef		20
	$M(AB)_3$		2
	M(ABA)cde		9
	$M(ABC)_2$		11
	M(ABBA)cd		7
	M(ABCBA)d		7

distinguished from their dipole moment values. Since the bond moments in the *trans*-isomer are equal and opposite, the net dipole moment becomes zero. The dipole moments for *cis*-isomers in such complexes are observed in up to 12 Debye units. Thus, a non-zero value of dipole moment in the case of square planar complexes of the type [Ma_2b_2] is an indication of *cis*-isomerism, while a *trans*-isomer is characterized by zero dipole moment.

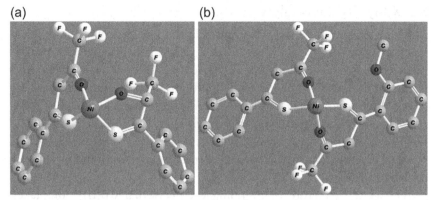

Figure 32 (a) *Cis-* and (b) *trans-*square planar complexes of monothiodiketone ligand.

A series of metal chelates based on monothiodiketone ligands are prepared and their geometrical orientations have been identified using dipole moment measurements. The square planar metal chelates of Ni(II), Pt(II), Pd(II) and Cu(II) showed the dipole moment values of 5.64, 6.15, 5.95 and 4.82 Debye units, respectively. These high values of dipole moment are clear indicators of the *cis-*isomerism in such complexes, as shown in Figure 32.

These monothiodiketone ligands also form octahedral complexes of the type M(AB)$_3$ and hence give *fac-* (*cis-*) and *mer-* (*trans-*) isomers. The isomer obtained upon synthesis of this Co(II) metal chelate shows a dipole moment of 6.75 Debye units and an assigned *fac-*configuration. The isomers of the complex are shown in Figure 33. This assignment is further supported by ^{1}H and ^{19}F nuclear magnetic resonance (NMR) spectroscopy [5].

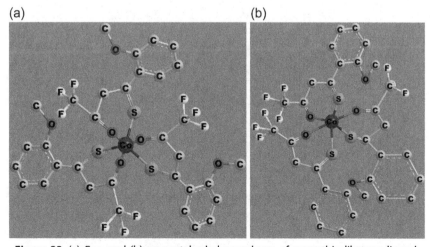

Figure 33 (a) *Fac-* and (b) *mer-*octahedral complexes of monothiodiketone ligand.

5.2 X-ray Crystal Analysis

X-ray crystal analysis is capable of providing more structural information than any other technique. With a single crystal, it is possible to solve the structure of complexes using X-ray diffraction. Since electrons diffract the X-rays, the centres of the electron clouds are located using this method. A computer program named Oak Ridge Thermal Ellipsoid Plot (ORTEP) gives the diagram of the complex, known as an ORTEP diagram. Using ORTEP diagrams, square planarity of several Pt(II) complexes has been confirmed. This arrangement has also been established for four coordinated complexes of Ag(II), Cu(II) and Au(III).

The crystal structure of mononuclear octahedral Co(III) complexes based on *N*-salicylidene-*o*-aminophenol and its derivatives has been proved using X-ray crystallography [6]. In a typical structure shown in Figure 34, the *trans*-angles of the octahedron are around 175°, while the *cis*-angles are in the range of 85.6–97.2°. Due to the meridional character of each ligand, two pairs of *trans*-coordination sites are each occupied by phenolate oxygen atoms of the same ligand group, whereas the third pair of coordination sites consists of nitrogen atoms from different ligands.

5.3 Infrared (IR) Spectroscopic Technique

The bands in the infrared spectra of the complexes are corresponding to the change in the dipole moment of the molecule. The vibrations that do not produce any change in the dipole moment do not show bands in the IR spectra. In the *trans*-isomers such as $[Co(NH_3)_4Cl_2]^+$ or $[Co(NH_3)_2Cl_2]^+$, the $Cl-M-Cl$ symmetrical stretching vibrations do not produce

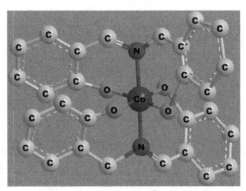

Figure 34 Co(III) complex based on *N*-salicylidene-*o*-aminophenol ligands.

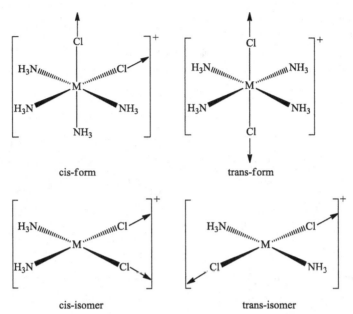

cis-form   trans-form

cis-isomer   trans-isomer

Figure 35 Dipole moments in *cis-* and *trans*-isomers.

any change in the dipole moment of the molecule, as shown in Figure 35. Hence, no band corresponding to this vibration is observed in IR spectra.

Whereas in the *cis*-isomers, the symmetrical, as well as unsymmetrical, stretching vibrations produce changes in the dipole moment. Thus, a *cis*-isomer shows a large number of bands due to Cl—M—Cl stretching. An infrared spectrum is thus useful in distinguishing the geometrical isomers of the complexes.

5.4 Grinberg's Method

In support of Werner's coordination theory, A. A. Grinberg developed a unique experimental method to distinguish the *cis-* and *trans*-isomers in coordination complexes. It is an important chemical method in which a chelating ligand with two donor atoms is allowed to react with *cis-* and *trans*-isomers separately. With *cis*-isomers, both the donor atoms of a chelating ligand coordinate to the central atom, forming a chelate with a five- or six-membered ring. While in the case of the *trans*-isomer, the chelating ligand can coordinate to the central atom by only one of the donor atoms. Here it functions as a monodentate ligand and forms an ordinary complex but not chelate.

Oxalic acid $(COOH)_2$, glycine (H_2N-CH_2-COOH) and ethyl-enediamine $(H_2N-CH_2-CH_2-NH_2)$ are the most frequent chelating ligands used in this method. Reactions between *cis-* and *trans-*isomers of $[Pt(NH_3)_2Cl_2]$ and oxalic acid, as well as glycine ligands, are shown in Figure 36.

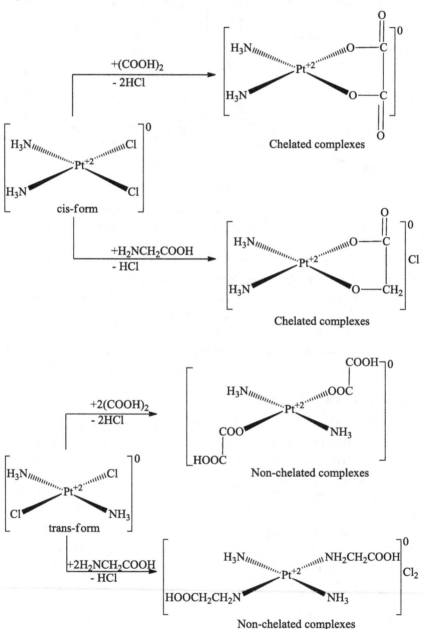

Figure 36 Reactions between isomers of $[Pt(NH_3)_2Cl_2]$ and oxalic acid/glycine.

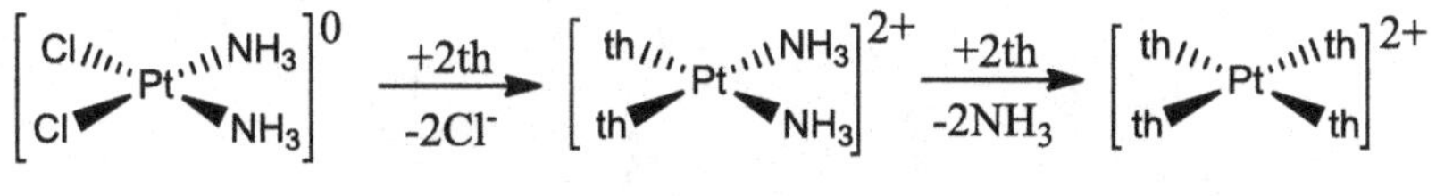

Figure 37 Reaction between *cis*-[Pt(NH$_3$)$_2$Cl$_2$] and thiourea.

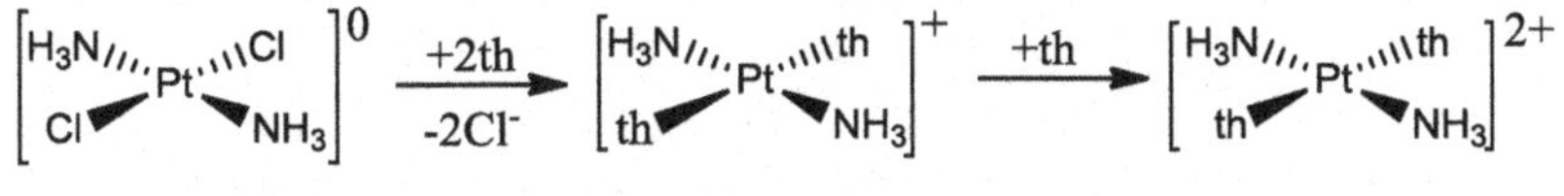

Figure 38 Reaction between *trans*-[Pt(NH$_3$)$_2$Cl$_2$] and thiourea.

5.5 Kurnakov's Method

This method is based on the *trans*-directing ability of various ligands. Kurnakov used the phenomena of *trans*-effect to distinguish the geometrical isomers of square planar complexes of [PtA$_2$X$_2$] type by treating them with thiourea (th). Here, Pt−N bond is stronger than Pt−Cl bond and the *trans*-directing ability of th is greater than NH$_3$.

While reacting a *cis*-isomer of [Pt(NH$_3$)$_2$Cl$_2$] with th, the two weakly bonded chloro ligands get replaced by thiourea in the first step. In the second step, since the thiourea has higher *trans*-directing ability than ammonia, they guide two more thiourea molecules to occupy *trans*-positions with respect to themselves. Thus, all the four ligands on the substrate get substituted by thiourea giving [Pt(th)$_4$], as shown in Figure 37.

While in case of the *trans*-isomer of [Pt(NH$_3$)$_2$Cl$_2$], the two weakly bonded chloro ligands are replaced by thiourea in the first step, but in the second step, the higher *trans*-effect of thiourea does not allow the ammine ligands to be sufficiently reactive so as to be replaced by the remaining thiourea ligands. Thus, the *trans*-isomer does not permit full substitution by thiourea and gives [Pt(NH$_3$)$_2$(th)$_2$], as shown in Figure 38.

In addition to thiourea, thiosulfate (S$_2$O$_3$)$^{-2}$ also shows similar reactions with the complexes of type [PtA$_2$X$_2$].

6. OPTICAL ISOMERISM OR MIRROR IMAGE ISOMERISM

In order to distinguish the optical isomerism visually without much theoretical discussion, consider Figure 39.

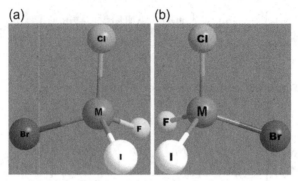

Figure 39 (a) Tetrahedral Mabcd complex and (b) its mirror image.

From Figure 39, it is seen that the complex is not superimposable on its mirror image. The molecules that are not superimposable on their mirror image are called chiral molecules. The reason behind the non-superimposability is the presence of asymmetric bonding on the central metal ion. This asymmetry is characterized by the attachment of all the four different ligands, viz, a, b, c and d to the centre. Chiral molecules do not have a plane of symmetry. If a molecule exists in two isomeric forms, which are mirror images of each other and are not superimposable over one another, the pair of such isomers is called enantiomers. When such enantiomers have an adequately long lifetime so that they can be separated, the phenomenon of optical isomerism occurs. These isomers are called optical isomers, as they are optically active. While one of the enantiomers rotates the plane of the polarized light in one direction by a particular angle, the other rotates it in the opposite direction with the same angle. The isomer that rotates the plane of the polarized light toward the right side (clockwise direction) is said to be dextrorotatory, and it is represented by a (+) sign. The isomer which rotates the plane of polarized light toward the left side (anticlockwise direction) is said to be laevorotatory, and it is represented by a (−) sign. However, just by looking at the structure, it is not possible to say whether a particular isomer will rotate the light in which direction. Hence, in absence of a large amount of experimental data and absolute necessity, the above-mentioned qualifications will be avoided in labelling the isomers. Instead, a convenient way of denoting the optical isomers in coordination complexes, using the prefixes 'Δ' (delta) and 'Λ' (lambda), will be employed in the further discussion. The prefix 'Δ' indicates the right-handedness in the structure, while the prefix 'Λ' corresponds to the left-handedness in the structure. Both isomers have exactly identical physical and chemical properties. A racemic mixture is not capable of rotating the plane-polarized light. It contains an exactly equal proportion of '+' and '−' forms. In a racemic mixture, one form rotates the plane of polarized light in one direction, which is balanced by the other form in the opposite direction.

6.1 Optical Isomerism: Four-Coordinate Complexes

6.1.1 Square Planar Complexes

Optical isomerism is not possible in any type of square planar complexes, as they possess at least one plane of symmetry (molecular plane). However, in one of the rare cases, a complex $[Pt(NH_2CH(C_6H_5)CH(C_6H_5)(NH_2)(NH_2CH_2C(CH_3)_2NH_2)]Cl_2$ was resolved into enantiomers, as shown in Figure 40. The platinum complex has a distorted square planar geometry, but it cannot be regarded as tetrahedral geometry.

This structure has no plane of symmetry and hence is unsymmetrical and optically active and gives an optical isomer.

6.1.2 Tetrahedral Complexes

Only an asymmetric tetrahedral molecule characterized by the lack of plane of symmetry [Mabcd] can show optical isomerism.

The Δ- and Λ-optical isomers of As^{+3} ion complex $[As(CH_3)(C_2H_5)S(C_6H_4COO)]^{+2}$ are shown in Figure 41.

Tetrahedral complexes of Be(II), B(III) and Zn(II) with asymmetrical bidentate ligands also exhibit optical isomerism. A complex shown in Figure 42, bis(benzoylacetonato)beryllium(II), $[Be(C_6H_5COCHCOCH_3)_2]$ is an important example of this class.

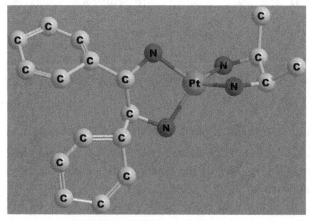

Figure 40 $[Pt(NH_2CH(Ph)CH(Ph)(NH_2)(NH_2CH_2C(CH_3)_2NH_2)]Cl_2$.

(a)   (b)

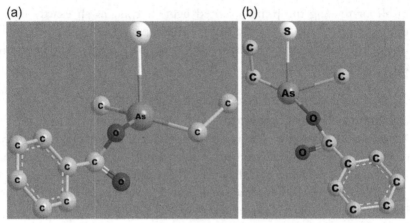

Figure 41 (a) Δ- and (b) Λ-isomers of [As(CH$_3$)(C$_2$H$_5$)S(C$_6$H$_4$COO)]$^{+2}$.

(a)

(b)

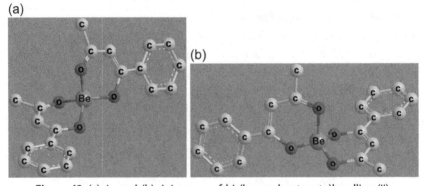

Figure 42 (a) Δ- and (b) Λ-isomers of bis(benzoylacetonato)beryllium(II).

In this complex, there is no centre or plane of symmetry, and the mirror image is not superimposable on itself. Thus, it will give Δ- and Λ-forms.

6.2 Optical Isomerism: Six-Coordinate Complexes

The octahedral geometry of the six-coordinate complexes yields not only the geometrical isomerism of the kind discussed earlier but also of mirror image isomerism leading to optical activity.

Optical activity is very common in the following types of octahedral complexes.

(a)   (b)

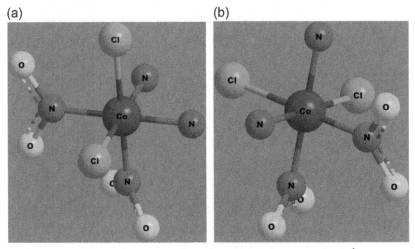

Figure 43 (a) Δ- and (b) Λ-isomers of $[Co(NH_3)_2Cl_2(NO_2)_2]^{-1}$.

6.2.1 Octahedral Complexes Containing Only Monodentate Ligands

6.2.1.1 $[Ma_2b_2c_2]^{\pm n}$ Type

This class is represented by a complex $[Co(NH_3)_2Cl_2(NO_2)_2]^{-1}$, which exists as two optical isomers, as shown in Figure 43.

6.2.1.2 $[Mabcdef]^{\pm n}$ Type

Only Pt(IV) complexes are known to form this type of complex. There are 15 geometrical isomers, each of which can exist in Δ- and Λ-forms (i.e. each of which a non-superimposable mirror image arises) to give a total of 30 isomers. For one form of $[Pt(py)(NH_3)(NO_2)(Cl)(Br)(I)]$, the optical isomers are shown in Figure 44.

(a)   (b)

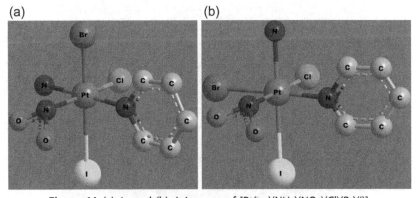

Figure 44 (a) Δ- and (b) Λ-isomers of $[Pt(py)(NH_3)(NO_2)(Cl)(Br)(I)]$.

(a)   (b)

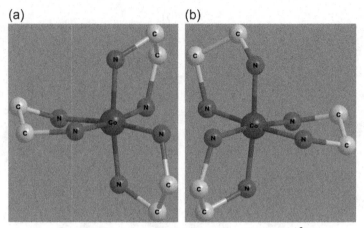

Figure 45 (a) Δ- and (b) Λ-isomers of $[Co(en)_3]^{+3}$.

6.2.1.3 $[M(AA)_3]^{\pm n}$ Type

Here, 'AA' is a symmetrical bidentate chelating ligand, which may be a neutral molecule or negative ions. The Δ- and Λ-isomers of such a complex, $[Co(en)_3]^{+3}$, are shown in Figure 45.

Other examples of this type of complex exhibiting optical isomerism are $[Co(pn)_3]^{+3}$, $[Pt(en)_3]^{+3}$, $[Cr(C_2O_4)_3]^{+3}$, $[Cd(pn)_3]^{+2}$ and $[Fe(C_2O_4)_3]^{-3}$.

6.2.1.4 $[M(AA)_2a_2]^{\pm n}$ Type

Here, 'AA' is the symmetrical bidentate chelating and 'a' is a monodentate ligand. The complex $[Co(en)_2Cl_2]^+$ has two geometrical isomers

(a)   (b)

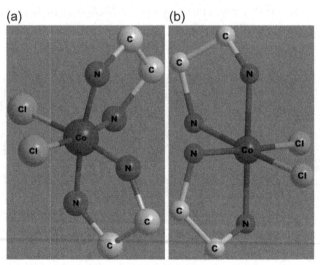

Figure 46 (a) Δ-*cis*- and (b) Λ-*cis*- isomers of $[Co(en)_2Cl_2]^+$.

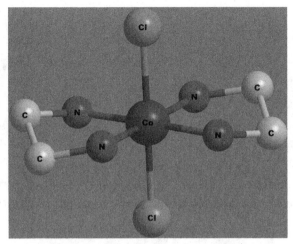

Figure 47 *Trans*-isomer of $[Co(en)_2Cl_2]^+$.

(i.e. *cis*–*trans*-isomers), as already discussed in the preceding section. In *cis*-isomers, there is no plane of symmetry, hence it shows optical isomerism, as shown in Figure 46.

In *trans*-isomers, as shown in the figure, there is a plane of symmetry, which makes its mirror image superimposable over itself. Hence it is optically inactive. We often come across such a member in the set of stereoisomers. Such form is known as '*meso*' form, as shown in Figure 47.

6.2.1.5 $[M(AA)_2ab]^{\pm n}$ Type

These complexes also exist in three forms, out of which two forms are optically active and a third form is inactive, as illustrated by $[Co(en)_2(NH_3)Cl]$ and shown in Figure 48.

(a)   (b)

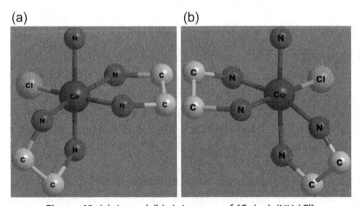

Figure 48 (a) Δ- and (b) Λ-isomers of $[Co(en)_2(NH_3)Cl]$.

(a)   (b)

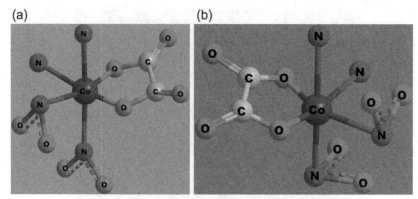

Figure 49 (a) Δ-*cis*- and (b) Λ-*cis*- isomers of $[Co(C_2O_4)(NH_3)_2(NO_2)_2]$.

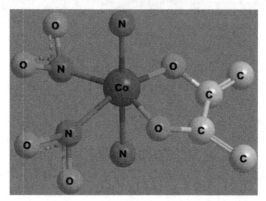

Figure 50 *Meso–trans*-isomer of $[Co(C_2O_4)(NH_3)_2(NO_2)_2]$.

6.2.1.6 $[M(AA)a_2b_2]^{\pm n}$ Type

The complexes with the above mentioned formula also exist in three forms, out of which two forms are optically active and a third one is a *meso*-form, as shown in Figures 49 and 50.

6.2.1.7 Octahedral Complexes Containing Optically Active Ligands

Chiral molecules in the role of ligand add a variety to the optical isomerism in metal complexes. 1,2-diaminopropane (pn) is an unsymmetrical biden-tate ligand possessing chiral carbon. The complex $[Co(en)(pn)(NO_2)_2]^+$ can form two *cis*- and two *trans*-isomers. Both the *cis*-forms give two optical isomers, as shown in Figures 51 and 52.

6.2.1.8 Octahedral Complexes Containing Polydentate Ligands

Ethylene diamine tetraacetate $(edta^{-4})$ is a popular hexadentate ligand in coordination chemistry.

(a)   (b)

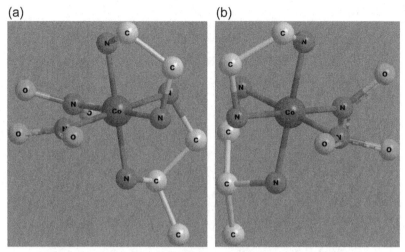

Figure 51 (a) Δ-*cis*- and (b) Λ-*cis*- isomers of [Co(en)(pn)(NO$_2$)$_2$]$^+$.

(a)   (b)

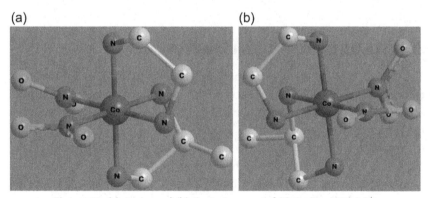

Figure 52 (a) Δ-*cis*- and (b) Λ-*cis*- isomers of [Co(en)(pn)(NO$_2$)$_2$]$^+$.

[Co(edta)]$^-$ exists in two optical isomers (Δ- and Λ-forms), as shown in Figure 53.

7. MAXIMUM NUMBER OF ENANTIOMERIC PAIRS IN DIFFERENT COMPLEXES

Owing to various geometries and a wide variety of ligands, a detailed treatise on all types of complexes is avoided. However, Table 4 can provide a guide to study isomerism in many other complexes.

(a)   (b)

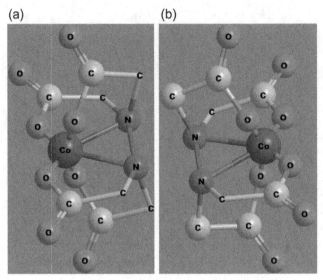

Figure 53 (a) Δ- and (b) Λ-isomers of [Co(edta)]⁻.

8. RESOLUTION OF RACEMIC MIXTURE

A mixture of (+) and (−) isomers in equal proportion is known as a racemic mixture. A racemic mixture is optically inactive, as expected. Sometimes, a racemic mixture exhibits different physical and chemical properties than any other isomer. The separation of a racemic mixture into its enantiomeric components ((+) and (−) forms) is called resolution. Since the (+) and (−) forms have the same physical and chemical properties, they cannot be separated by ordinary methods such as fractional crystallization and fractional distillation. The use of enzymes and chromatography has enabled the separation of some isomers.

Diastereoisomers are optically active isomers that are not enantiomers. The *cis-* and *trans-*isomers of a complex can sometimes be diastereoisomers. An important method in the resolution of a racemic mixture is using the formation of diastereomers.

If enantiomers of a (+) and (−) mixture are acid, they give salt upon treatment with an optically active base. In the same way, basic enantiomers of a racemic mixture give salts upon reaction with an optically active acid. The reactions with an optically active acid/base, as required, give two diastereomeric salts. These diastereomers can be separated by fractional crystallization on the basis of the difference in their solubility in water.

Table 4 Maximum number of enantiomeric pairs in different complexes

Sr. No.	Formula	Geometry	Number enantiomeric pairs
1.	Ma_4	Tetrahedral	0
	Mabcd		1
2.	Ma_4	Square planar	0
	Ma_3b		0
	Ma_2b_2		0
	Ma_2bc		0
	Mabcd		0
	$M(AA)_2$		0
	$M(AB)_2$		0
	$M_2A_2B_4$		0
3.	Ma_6	Octahedral	0
	Ma_5b		0
	Ma_4b_2		0
	Ma_3b_3		0
	Ma_4bc		0
	Ma_3bcd		1
	Ma_2bcde		6
	Mabcdef		15
	$Ma_2b_2c_2$		1
	Ma_2b_2cd		2
	Ma_3b_2c		0
	$M(AA)_2a_2$		1
	$M(AA)_2ab$		1
	$M(AA)a_2b_2$		1
	$M(AA)(BC)de$		5
	$M(AB)(AB)cd$		5
	$M(AB)(CD)ef$		10
	$M(AB)_3$		2
	$M(ABA)cde$		3
	$M(ABC)_2$		5
	$M(ABBA)cd$		3
	$M(ABCBA)d$		3

The separated salts are treated with an optically active material to regenerate the optically active (+) and (−) forms.

For the resolution of (+)-cis-[Co(en)$_3$]Cl$_3$ and (−)-cis-[Co(en)$_3$]Cl$_3$, the racemic mixture is treated with an optically active (+)-tartaric acid. Two of Cl$^-$ ions of the complexes are replaced by (+)-tartaric ion to give two diastereomers as shown below.

(+)(-)[Co(en)$_3$]Cl$_3$ + (+)-tartaric acid → (+)-[Co(en)$_3$]$^{+3}$Cl$^-$((+)-tart)$^{-2}$ + (−)-[Co(en)$_3$]$^{+3}$Cl$^-$((+)-tart)$^{-2}$ + 4Cl$^-$. Upon crystallization, the $(+)$-cis-$[Co(en)_3]^{2+}Cl((+)$-tart$)^{2-}$ separates in the form of large crystals. Due to higher solubility, the crystals of $(-)$-isomers are obtained much later.

After separation, a treatment with concentrated hydrochloric acid gives the $(+)$ and $(-)$ chloride complexes.

$$(+)\text{-}[Co(en)_3]^{+3}Cl^{-1}((+)\text{-tart})^{-2} + 2HCl \rightarrow (+)\text{-}[Co(en)_3]Cl_3 +$$
$$(+)\text{-tartaric acid}$$

$$(-)\text{-}[Co(en)_3]^{+3}Cl^{-1}((+)\text{-tart})^{-2} + 2HCl \rightarrow (-)\text{-}cis[Co(en)_3]Cl_3 +$$
$$(+)\text{-tartaric acid}$$

Numerous resolving agents like antimonyl tartarate ion (SbOd-tart$^-$) and $(+)$-α-bromo camphor-π-sulphonate anion have been used for resolution of cationic complexes. Resolving agents used for anionic complexes are obtained from bases like brucine and strychnine. Methods that are more sophisticated are required for the resolution of neutral complexes. The neutral complexes can be resolved by preferential adsorption on quartz or sugars or extraction using asymmetric solvents. They can also be resolved using gas chromatography with the capillary packed with solvents (Golay columns).

9. EXERCISES

9.1 Multiple Choice Questions

N.B.: More than one option can be correct

1. Which of the following complex ions shows geometrical isomerism?
   (a) $[Cr(H_2O)_4Cl_2]^+$ (b) $[Pt(NH_3)_3Cl]$
   (c) $[Co(NH_3)_6]^{3+}$ (d) $[Co(CN)_5(NC)]^{3-}$

2. The complexes of the type $[Pd(C_6H_5)_2(SCN)_2]$ and $[Pd(C_6H_5)_2(NCS)_2]$ are _____isomers.
   (a) linkage   (b) coordination
   (c) ionization   (d) geometrical

3. The complexes $[Co(SO_4)(NH_3)_5]Br$ and $[Co(SO_4)(NH_3)_5]Cl$ represent_____isomerism.
   (a) linkage   (b) ionization
   (c) coordination   (d) no

4. _____ type of isomerism exists between $[Cr(H_2O)_6]Cl_3$ (violet) and $[Cr(H_2O)_5Cl]Cl_2 \cdot H_2O$ (green).

   (a) linkage   (b) solvate
   (c) ionization   (d) coordination

5. Identify the optically active compounds from the following:

   (a) $[Co(en)_3]^{3+}$ (b) trans-$[Co(en)_2Cl_2]^+$
   (c) cis-$[Co(en)_2Cl_2]^+$ (d) $[Cr (NH_3)_5Cl]$

6. Identify the complexes that show linkage isomerism from the following:

   (a) $[Co(NH_3)_5(NO_2)]^{2+}$ (b) $[Co(H_2O)_5CO]^{3+}$
   (c) $[Cr(NH_3)_5SCN]^{2+}$ (d) $[Fe(en)_2Cl_2]^+$

7. $K_3[Fe(CN)_6]$ and $K_4[Fe(CN)_6]$ are _____ isomers.

   (a) linkage   (b) not
   (c) ionization   (d) coordination

9.2 Short/Long Answer Questions

1. Which one of the following complexes is expected to be optically active?

   (a) cis-$[CrCl_2(ox)_2]^{3-}$ (b) trans-$[CrCl_2(ox)_2]^{3-}$

2. Which types of isomerism is exhibited by the following complexes?

   (i) $K[Cr(H_2O)_2(C_2O_4)_2]$ (ii) $[Co(en)_3]Cl_3$
   (iii) $[Co(NH_3)_5(NO_2)](NO_3)_2$ (iv) $[Pt(NH_3)(H_2O)Cl_2]$

3. Justify that $[Co(NH_3)_5Cl]SO_4$ and $[Co(NH_3)_5SO_4]Cl$ are ionization isomers.

4. How many geometrical isomers are possible in the following two complexes?

   (i) $[Cr(C_2O_4)_3]^{-3}$ (ii) $[Co(NH_3)_3Cl_3]$

5. Sketch the structures of optical isomers of the following complexes:

   (i) $[Cr(C_2O_4)_3]^{-3}$ (ii) $[PtCl_2(en)_2]^{+2}$
   (iii) $[Cr(NH_3)_2Cl_2(en)]^+$

6. Sketch all of the isomers of the following complexes:

   (i) $[CoCl_2(en)_2]^+$ (ii) $[Co(NH_3)Cl(en)_2]^{+2}$
   (iii) $[Co(NH_3)_2Cl_2(en)]^+$

7. List all of the possible geometrical isomers of $[Pt(NH_3)(Br)(Cl)(py)]$. How many of these will exhibit optical isomerism?

8. A complex of the type $[M(AA)_2X_2]^{n+}$ is optically active; giving a suitable example, comment on the structure of such complex.

9. A complex $Co(NH_3)_2(H_2O)_2Cl_2Br_2$ exists in two isomeric forms. Upon treatment with silver nitrate, one of the forms gives 2 mol of silver bromide, while the other gives only 1 mol. Predict the structures of both the isomers.

10. A complex $Co(NH_3)_5(SO_4)Br$ exists in two isomeric forms. Upon treatment with silver nitrate, the first one gives silver bromide, while the second one does not. However, the second isomer gives precipitates with an aqueous solution of $BaCl_2$. Predict the structures of both the isomers.

11. Establish the co-ligand isomer for the complex cis-$[Pt(N_3)(CH_2 C_6H_4CN)(PPh_3)_2]$.

12. A, B and C are three isomers of a complex with the empirical formula $H_{12}O_6Cl_3Cr$ having chloro and aquo ligands. The isomer A does not lose any weight with H_2SO_4, while B and C lose 6.75% and 13.5% of their weight. Identify A, B and C.

13. A complex cation, $[Pt(NH_3)(NH_2OH)(NO_2)(C_6H_5N)]^+$, is optically active. Predict the geometry of the complex.

SOURCES FOR FURTHER READINGS

The topics discussed in this chapter are a part of a standard graduate curriculum. A majority of the textbooks with titles related to coordination chemistry can act as a source of further reading. Moreover, there are several web resources useful for further learning. Some of them are listed below:

http://wwwchem.uwimona.edu.jm/courses/IC10Kiso.html
http://www.chem.purdue.edu/gchelp/cchem/
http://en.wikipedia.org/wiki/Coordination_complex
http://nptel.ac.in/courses/104106064/

REFERENCES

Attempts have been made to include some illustrations from the research articles listed below:

[1] Hvastijová M, et al. Nucleophilic additions to pseudohalides in the coordination sphere of transition metal ions and coligand isomerism. Coord Chem Rev 1998; 175(1):17–42.

[2] Kato M, et al. Linkage isomerism of pentaammine(dimethylsulfoxide)ruthenium(II/III) complexes: a theoretical study. Inorg Chim Acta 2009;362(4):1199–203.

[3] Fehlhammer WP, Stolzenberg H. Isocyanide- and heteroallene-bridged metal complexes. III. Reactions of Pd(η^2-CS$_2$)(PPh$_3$)$_2$ with metal compounds. 'Coordination isomerism' in metallodithiocarboxylato metal complexes. Inorg Chim Acta 1980;44(0):L151–2.

[4] Artem'ev AV, et al. DFT study and dynamic NMR evidence for cis-trans conformational isomerism in square planar Ni(II) thioselenophosphinate, Ni(SeSPPh$_2$)2. J Organomet Chem 2014;768(0):151–6.

[5] Das M. Dipole moments of some transition metal complexes of five new monothio-β-diketones. Inorg Chim Acta 1982;60(0):165–9.

[6] Alexopoulou KI, et al. Mononuclear anionic octahedral cobalt(III) complexes based on N-salicylidene-o-aminophenol and its derivatives: synthetic, structural and spectroscopic studies. Spectrochim Acta Mol Biomol Spectrosc 2015;136(Part A):122–30.

CHAPTER 4

Thermodynamics and Kinetics of Complex Formation

Contents

Essentials of Coordination Chemistry
http://dx.doi.org/10.1016/B978-0-12-803895-6.00004-5

1. INTRODUCTION

The thermodynamic and kinetic studies are of paramount importance while synthesizing and reacting compounds. Using the knowledge of thermodynamic properties, it is possible to know the ease of a particular transformation. The value of standard state free energy of a reaction (ΔG^0) provides an insight to the spontaneity of a reaction. The reactions for which this value is negative is an easily possible reaction, but one needs to remember that 'impossible is nothing'; the reactions with positive values of ΔG^0 are difficult but not impossible. Once thermodynamics okays the transformation, it is the kinetics that gives us information about the speed of transformation.

2. STABILITY AND LABILITY OF COMPLEXES

The term stability generally indicates the ability of a complex to exist for a period of time. Thus, stable complexes can be stored for a long time and under suitable conditions. This is a relative term, as different complexes may exhibit variable extent of stability with respect to different reagents. Moreover, the stability of the complexes is also evaluated with reference to the action of heat or light on a compound. Thus, stability is a general term, and it may not be very useful unless used with suitable qualifications. While studying the reactions involving complex formation, both thermodynamic and kinetic stabilities require consideration.

Thermodynamic stability of a complex is a measure of the extent of formation or transformation of a complex at the stage of equilibrium. Thus, a thermodynamically stable species does not readily undergo any reaction, whereas the complexes that are unstable in thermodynamic sense are expected to be reactive. The basis of the thermodynamic stability of a complex is ΔG, which is the free energy change involved in the complex formation. The free energy change may be equated as $\Delta G = -RT \ln K = \Delta H - T\Delta S$, where ΔH

represents the enthalpy and ΔS represents the entropy change. In absence of the availability of ΔG values, the ΔH values also provide significant information regarding the stability of chemical entities. One needs to keep in mind that the thermodynamic stability of a complex does not give any information regarding the speed of transformation.

Kinetic stability refers to the speed with which transformations leading to the attainment of equilibrium take place. The complexes that react quickly are called labile complexes, while the slow-reacting complexes are named as inert complexes.

If the reactant species for a proposed reaction is thermodynamically stable, it will not undergo transformation without the application of forcing conditions. If the reactant is thermodynamically unstable, it can be considered as a green signal for a reaction to proceed. Kinetics at this stage must be referred to know the rate at which the proposed transformation will take place. The kinetic stability of a complex can also be evaluated from the ΔH values of the substitution reactions on a complex. The reactions selected for such study are generally the ones involving the exchange of water molecules. From the ΔH values for such reactions occurring in different metals, a qualitative conclusion regarding the lability of the metal complexes can be derived.

3. STEPS INVOLVED IN FORMATION OF A COMPLEX ML_n, STEPWISE AND OVERALL STABILITY CONSTANTS

The complex formation in a solution proceeds by the stepwise addition of the ligands to the metal ion. Consider the formation of a complex ML_n occurring through a series of steps as the following:

$$M + L \overset{K_1}{\rightleftharpoons} ML \quad K_1 = \frac{[ML]}{[M][L]}$$

$$ML + L \overset{K_2}{\rightleftharpoons} ML_2 \quad K_2 = \frac{[ML_2]}{[ML][L]}$$

$$ML_2 + L \overset{K_3}{\rightleftharpoons} ML_2 \quad K_3 = \frac{[ML_3]}{[ML_2][L]}$$

$$\vdots$$

$$ML_{n-1} + L \overset{K_n}{\rightleftharpoons} ML_n \quad K_n = \frac{[ML_n]}{[ML_{n-1}][L]}$$

The equilibrium constants K_1, K_2...K_n are stepwise formation constants, or stepwise stability constants, which indicate the extent of formation of different species corresponding to a particular step. The values of stepwise stability constants for a complex formation reaction mostly decrease successively $(K_1 > K_2 > K_3 > ... > K_n)$. A steady decrease in the values of these constants is due to a decrease in coordinated water ligands that are available to fresh ligands for replacement. Other than this, the decrease in the ability of metal ions with progressive intake of ligands, steric hindrance and columbic factors also contributes to a steady decrease in successive stepwise stability constant values.

The stepwise stability constants fail to include the information regarding the previous steps. In order to include the extent of formation of a species up to a particular step, the overall stability constants (β) are introduced. For instance, β_3 indicates the extent of formation up to step 3. Mathematically, it is the product of stepwise stability constants of steps 1 to 3.

$$M + 3L \overset{\beta_3}{\rightleftarrows} ML_3 \quad \beta_3 = \frac{[ML_3]}{[M][L]^3}$$

The equality of the product of stepwise stability constants K_1, K_2, K_3 and overall stability constant can be shown as the following:

$$K_1 \times K_2 \times K_3 = \frac{[ML]}{[M][L]} \times \frac{[ML_2]}{[ML][L]} \times \frac{[ML_3]}{[ML_2][L]} = \frac{[ML_3]}{[M][L]^3} = \beta_3$$

This relation can be generalized as $\beta_n = K_1 \times K_2 \times K_3 \times ... \times K_n$

Since the equilibrium constants are commonly expressed in logarithmic form, the equation for β_n can also be shown as $\log\beta_n = \sum_{n=1}^{n=n}\log K_n$

To illustrate this, consider the formation of $[Cu(NH_3)_4]^{+2}$ ion

$$Cu^{2+} + NH_3 \rightarrow [Cu(NH_3)]^{+2} \qquad K_1 = \left[Cu(NH_3)_2^{+2}\right]/[Cu^{2+}][NH_3]$$

$$[Cu(NH_3)]^{+2} + NH_3 \rightarrow \left[Cu(NH_3)_2\right]^{+2} \quad K_2 = \left[Cu(NH_3)_2^{+2}\right]/[Cu(NH_3)][NH_3] \text{etc.}$$

where K_1, K_2 are the stepwise stability constants and overall stability constant.

Also $\beta_4 = [Cu(NH_3)_4]^{+2}/[Cu^{2+}][NH_3]^4$

The addition of the four amine groups to copper shows a pattern in which the successive stability constants decrease. In this case, the four constants are

$\log K_1 = 4.0$, $\log K_2 = 3.2$, $\log K_3 = 2.7$, $\log K_4 = 2.0$.

Thus, $\log \beta_4 = 11.9$.

The instability constant or the dissociation constant of coordination compounds is defined as the reciprocal of the formation constant.

4. EXPLANATION OF LABILITY AND INERTNESS OF OCTAHEDRAL COMPLEXES

4.1 Valence Bond Theory (VBT)

According to VBT, octahedral complexes may be either outer-orbital complexes involving sp^3d^2 hybridization or inner-orbital complexes resulting from d^2sp^3 hybridization. In both the cases, e_g orbitals are involved in the hybridization.

As per VBT, the outer-orbital octahedral complexes are generally labile due to the weakness of the bonds of sp^3d^2 type, as compared with d^2sp^3 bonds. This prediction has been supported by complexes of Mn^{+2} ($3d^5$), Fe^{+3} ($3d^5$), Co^{2+} ($3d^7$), Ni ($3d^8$) and Cu^{2+} ($3d^9$).

In the case of the inner-orbital octahedral complexes, the six d^2sp^3 hybrid orbitals are occupied by six electron pairs from ligands. The electrons of the central metal occupy the t_{2g} orbitals. The inner orbital octahedral complexes which contain at least one d-orbital of t_{2g} set empty are known to be labile. Probably, this empty d-orbital is used to accept the electron pair from the incoming ligand during the formation of an activated complex with a higher coordination number. In the inert inner-orbital octahedral complexes, every d-orbital of t_{2g} set contains at least one electron.

4.2 Crystal Field Theory (CFT)

Substitution reactions in octahedral complexes follow either substitution nucleophilic unimolecular (SN [1]) or substitution nucleophilic bimolecular mechanism. These pathways involve five coordinated (trigonal bipyramidal/square pyramidal) and seven coordinated (pentagonal bipyramidal) intermediates, respectively. The formation of any of these intermediates involves the lowering of the symmetry. This, in turn, reduces the crystal field stabilization energy (CFSE). This loss in CFSE is the activation energy (E_a) required for the transformation. The octahedral complexes formed by the ions, for which there is a large loss in CFSE, do not react rapidly and are inert. However, the complexes of the ions for which there is little or no loss in CFSE are labile.

5. FACTORS AFFECTING THE STABILITY AND LABILITY OF COMPLEXES

5.1 Factors Affecting the Kinetic Stability of Complexes

The observed kinetic stability (lability or inertness) of a species is the result of multifaceted interplay between several factors. The effect of some such factors is summarized in Table 1.

5.2 Factors Affecting the Thermodynamic Stability of Complexes

A number of factors affect the stability of the complexes. Some of these factors depend on the nature of the central ion, while others depend on the nature of the ligand. A few such factors and their effects are discussed in Table 2.

6. BASES FOR DETECTION OF COMPLEX FORMATION

A complexed metal ion exhibits strikingly different properties, as compared to a free metal ion. The complex formation can be detected by studying such properties of metal ion. In addition to this, any property of a system, which is expected to vary with a change in the concentration of one of the species involved in the formation of the complex, can be used to sense the formation of a complex. Changes in properties like colour, solubility, melting point, stability towards oxidation or reduction, magnitude of ionic charge, solubility and crystalline form of the salts are also indicative of the formation of a complex. By careful consideration, it is possible to find one or more suitable methods for the detection of all types of complex formation in the solution.

Table 1 Effect of some factors on lability of complexes

Factor	Comments
Charge on the central metal ion	Lability order for some isoelectronic species is $[AlF_6]^{3-} > [SiF_6]^{2-} > [PF_6]^- > [SF_6]^0$.
Radius of the central metal ion	Order of lability for some [complexes](radius): $[Mg(H_2O)_6]^{2+}(0.65 \text{ Å}) < [Ca(H_2O)_6]^{2+}(0.99 \text{ Å}) < [Sr(H_2O)_6]^{2+}(1.13 \text{ Å})$.
Coordination number	Tetrahedral and square planar complexes exhibit higher lability towards isotopic ligand exchange compared to the octahedral moieties. This is because of the availability of a larger room for entry of the incoming ligands in case of four-coordinated complexes.

Table 2 Effect of some factors on the stability of complexes

Sr.No.	Factor	Comments
Central metal ion		
1.	Charge	The value of stability constant for $[Fe^{+3}(CN)_6]^{-3}$ is about four times greater than that of $[Fe^{+2}(CN)_6]^{-4}$.
2.	Radius	The hydroxide of magnesium (0.65 Å) has four times greater value of stability constant as compared to the hydroxide of calcium (0.99 Å).
3.	Crystal field stabilization energy (CFSE)	For a reaction $M + 6L \overset{\beta}{\rightleftharpoons} ML_6 \quad \beta = \dfrac{[ML_6]}{[M][L]^6}$ The overall stability constant follows the trend as $d^0 < d^1 < d^2 < d^3 > d^4$ and $d^5 < d^6 < d^7 < d^8 > d^9 > d^{10}$
	Class	
4.	Class a → stable complexes with ligands having the coordinating atoms of the second period elements (e.g. N, O, F)	Class a metals: Na, Al, Ca, Ti, Fe and lanthanides
5.	Class b → stable complexes with ligands having the coordinating atoms from third period of elements (e.g. P, S, Cl)	Class b metals: Rh, Pd, Ag, Ir, Pt, Au and Hg
Ligand		
6.	Ratio of charge to size	Greater negative charge coupled with smaller size gives higher value of the ratio. Thus, the stability constant of a fluoro complex (10^6) of ferric ion is much higher than that of corresponding chloro complex (20).

Continued

Table 2 Effect of some factors on the stability of complexes—cont'd

Sr.No.	Factor	Comments
7.	Basicity	Stronger Lewis bases readily donate electron pairs and form complexes with greater stability. In line of this, ammonia forms more stable complexes than water, which in turn forms more stable complexes than hydrofluoric acid.
8.	Chelate effect	Complexes containing chelate rings are generally more stable than those without rings. The value of stability constant for an ammonia complex with Ni(II) is 7.98, whereas with ethylenediamine (en) is 18.2. The differences in entropy between chelate and non-chelate complex reactions essentially cause this change in the values of stability constants. The formation of chelate results in greater disorder as it involves the formation of a larger number of free particles in the products, while there is no change in the number of particles during the formation of analogous non-chelated complexes. The chelates are formed when a polydentate ligand uses more than one of its coordinating atoms to form a ring involving the metal ion. Here, the dissociation of one of the bonds with the metal ion does not detach the ligand molecule, and hence, the possibility of reestablishment of this bond is more, as compared to an ordinary complex. Thus, chelation provides additional stability to a complex molecule.
9.	Number of chelate rings	Ethylenediamine (en) and diethylenetetramine (dien) are capable of forming one and two chelate rings, respectively. The 1:2 complexes of these ligands with Fe^{+2} ion register the values of stability constants of 7.7 and 10.4, respectively. The chelates containing five- and six-membered chelate rings show maximum stability.
10.	Steric hindrance	The Ni(II) complex of oxine (8-hydroxy quinolone) has higher stability constant value of 21.8, as compared to an analogous complex with 2-methyl-8-hydroxy quinolone (17.8)

6.1 Colour Change

Many a times a colour change is accompanied with the complex formation [1]. When ammonia is added to Cu^{2+} ion, a deep colour is obtained due to the formation of $[Cu(NH_3)_4]^{2+}$, a complex ion. Similarly, pink Co^{+2} and green Fe^{+2} turn blue and yellow, respectively, under the influence of Cl^- and CN^- ligands to form the complex ions $[CoCl_4]^{2-}$ and $[Fe(CN)_6]^{4-}$. These colour changes give an indication of the complex formation.

6.2 Solubility

In certain cases, the formation of a complex increases the solubility of a sparingly soluble salt. When potassium cyanide is added to a solution of sparingly soluble silver cyanide, solubility of silver cyanide increases due to complex formation.

$$AgCN + KCN \rightleftarrows K\left[Ag(CN)_2\right]$$
$$K\left[Ag(CN)_2\right] \rightleftarrows K^+ + \left[Ag(CN)_2\right]^-$$

A similar change in solubility is observed when ammonia is added to a solution of sparingly soluble silver chloride. The formation of complex $[Ag(NH_3)_2]Cl$, which ionizes as complex cation and chloride ion, causes an apparent change in the solubility of silver chloride.

$$AgCl + 2NH_3 \rightleftarrows \left[Ag(NH_3)_2\right]Cl$$

In the case of nonionic complexes, the solubility of the complex in a water immiscible organic solvent and its distribution coefficient can be used as a measure of the extent of complex formation.

6.3 Change in Chemical Properties

The fact that a complexed metal ion reacts differently than a free ion allows us to distinguish them and, in turn, can give evidences of complex formation. The addition of hydroxide ion to a solution of Fe^{+2} in the presence of a disodium salt of ethylenediaminetetraacetic acid (EDTA) does not result into precipitation of $Fe(OH)_2$. This is because iron is in the form of complex $[Fe(EDTA)]^{2-}$. Similarly, Cl^- does not cause precipitation of Ag^+ from aqueous ammonia solution, as Ag^+ ions are present in the form of complex $[Ag(NH_3)_2]^+$.

However, the formation of complexes with low thermodynamic stability fails to qualify such a test. This is due to an appreciable concentration of free metal ions in the solution. The complexes between ammonia and Ca^{2+}, Zn^{2+} or Al^{3+} give all the usual precipitation reactions of the free metal ion.

6.4 pH

Most of the complex formation reactions involve loss of one or more protons from the ligands. Thus, the complex formation may be considered to be the displacement of one or more, usually, weak acidic protons of the ligand by the metal ion. This displacement increases [H^+] ion concentration in the medium, resulting in a drop in the pH. Thus, the pH effect can be used as an indication of the extent of complex formation.

6.5 Freezing Point

The properties that depend on the number of particles present in a solution but are independent of their nature are called colligative properties. The depression of freezing point is a colligative property. Since the complex formation is a process, in which the number of particles decrease, a depression in freezing point is expected upon complexation.

6.6 Electrical Conductivity

Electrical conductance in a solution depends on the concentration of ions with high mobility. Complex formation is often accompanied by the disappearance or formation of ions with different mobility. Thus, a change in the electrical conductance of the solution is usually observed, which can be used as a basis for the detection of the formation of a complex [2].

The formation and disappearance of ions is demonstrated by the following example:

$$\left[Cu(H_2O)_X\right]^{2+} + 2(CH_3COO)^- + 2NH_2CH_2COOH \rightarrow$$
$$\left[Cu(NH_2CH_2COO)_2\right] + CH_3COOH$$

If the complex is sufficiently stable, this method can also be used to determine the charge on the complex by measuring the molar conductivity at infinite dilution. Sometimes, it is also possible to determine the structure of the complex, provided the number of ions and their charges are known.

6.7 Ion Exchange Adsorption

This method is very useful in a system where the charge on the complex is different from that of the metal ion [3,4]. An anionic exchange resin cannot adsorb Zn^{+2} cation from an aqueous solution without the presence of a large excess of hydrochloric acid. In the presence of hydrochloric acid, the zinc is in the form of negative ion $[ZnCl_4]^{2-}$, which can be adsorbed on an anion exchange resin. Comparison of the concentrations of zinc in the resin and aqueous phase gives an insight into the extent of complex formation.

6.8 Dipole Moment

A bond between two different atoms cannot be purely covalent. Depending on the electronegativity of the bonded atoms, the bonded electron pair shifts toward one of the atoms. A vector showing the magnitude and direction of the shifting of this electron pair is known as a bond moment. In polyatomic molecules, each bond has an individual bond moment. The dipole moment (μ) of a molecule is the vectorial addition of such bond moments. In the cases where all the individual bond moments are zero (homoatomic molecules), the resultant dipole moment is always zero. However, the converse is not true. Because of equal and opposite values of individual bond moments, some molecules show a zero net dipole moment. As applied to coordination compounds, a dipole measurement can be of value in distinguishing between isomers of a compound, particularly between *cis-* and *trans-* isomers. A *trans-* isomer exhibits a low or zero dipole moment.

6.9 Magnetic Measurements

Magneto chemistry is a very powerful tool in the hands of a synthetic coordination chemist. The magnetic behaviour reveals the number of unpaired electrons in a complex. This study also provides the information regarding the valence of the metal ion, the bond type and in turn, the geometry of the coordination compounds.

The magnetic moment (μ) of the transition metal ion in their compounds is directly related to the number of unpaired electrons (n) and can be equated as $\mu = \sqrt{[n(n+2)]}$. In some cases, where inner d-orbitals are used for hybridization, the complex formation may result in a decrease of n. A decrease in the value of a magnetic moment becomes an evidence of complexation in some cases. A free Ni^{+2} $(Z = 28)$ has eight electrons in five 3d orbitals. Here, six electrons are paired while two remain unpaired, exhibiting a magnetic moment of 2.83 bohr magneton (BM). In a square planar complex of Ni(II), the dsp [2] hybridization requires one of the d-orbitals to be vacated, which results into the pairing to two unpaired electrons. Thus, a square planar complex of Ni(II) is diamagnetic $(\mu = 0)$.

6.10 Visible Absorption Spectra

The visible absorption spectroscopy, also known as electronic spectroscopy, is of dominant importance in coordination chemistry. The absorption of light in the visible region (330–800 nm) by a metal ion depends on the electronic transitions within the ion. It also depends on the type and strength of metal-ligand bonding.

Thus, the metal ions containing incompletely filled d-orbitals give different colours, depending on the nature of the ligands attached with the metal ion. In most cases, the intensity of absorption is increased, and the absorption shifts to a lower wavelength when coordinated water molecules are replaced by ligands more basic than water.

6.11 Nuclear Magnetic Resonance (NMR)

Nuclear magnetic resonance spectroscopy, commonly known as NMR spectroscopy, is a technique that exploits the magnetic properties of certain atomic nuclei to determine physical and chemical properties of atoms, and in turn, the molecules in which they are contained. It is based on nuclear magnetic resonance and can provide detailed information about the structure of molecules. It is applicable to a sample containing nuclei with resultant nuclear spin. When such a nucleus is placed in a strong magnetic field, transitions occur between different spin energy states of the nucleus due to the absorption of frequencies.

In addition to proton NMR, ^{13}C and ^{19}F NMR are useful in studying metal ligand bonding in hydrido and fluoro complexes and of ligands having hydrogen, e.g. acetylacetonate (acac).

The chemical shift (designated as δ and having unit ppm) in the proton NMR provides information about the complex formation. Here also, the method can be utilized, even if there is no change in the number of unpaired electrons of the central atom or ion on complex formation.

6.12 Electron Spin Resonance (ESR) Spectra

Electron spin resonance (ESR) spectroscopy, also known as electron paramagnetic resonance (EPR) spectroscopy, is analogous to NMR spectroscopy. It is 'an electronic version' of NMR spectroscopy. This technique is very useful in studying the complexes containing unpaired electrons. It gives information about the distribution of unpaired electrons in the molecule, which in turn helps in mapping the extent of delocalization of electrons over the ligand. When placed in a magnetic field, the molecules with unpaired electrons undergo transitions between different states. The absorption takes place in the microwave region and the energy of transition E is given by $E = h\nu = g\beta H$

where β is the magnetic moment expressed in Bohr Magnetons, H is the strength of the field and g is the ratio of magnetic moment to angular momentum, known as gyromagnetic ratio.

These measurements provide the value of magnetic moment. The sub-bands in the absorption bands can explain the nature of the metal-ligand bond in the complex. The ESR measurements require a very small quantity of a sample. It is particularly useful for studying complex formation, where there is no change in the number of electrons of the central metal or ion.

6.13 Electromigration

The formation of negatively charged complex ions can be detected using electrophoresis and electrodialysis. In both these processes, the charged species are forced to move by application of external electric field. While the electrophoresis is applicable to dispersed particles with a surface charge, electrodialysis is used to transport the ions in the solution. The reversed movement of complexed metal ion (coloured ion) toward the positive electrode instead of negative electrode serves as an indicative of complex formation.

Consider the formation Cobalt(III) aspartic acid complex. The complex shown in Figure 1 is an anionic complex. The movement of free Co(III) ion under the electric field will be reversed by the event of complex formation.

6.14 Change in Reduction Potential

The reduction potential of a species is its tendency to gain electrons and get reduced. It is measured in millivolts or volts. Larger positive values of reduction potential are indicative of a greater tendency to get reduced. We have seen earlier that the complex formation is accompanied by a decrease in the ionic activity of the metal. Therefore, the reduction potential of the metal ion decreases upon complexation.

The standard reduction potential for a free Co^{+3} ion is 1.853 V, while in complexed state $[Co(NH_3)_6]^{+3}$, it decreases to 0.1 V. Similarly, a free Fe^{+3} ion has a standard reduction potential of 0.771 V, but $[Fe(CN)_6]^{-3}$ has the value of 0.36 V.

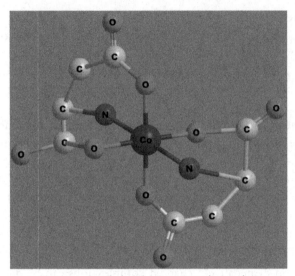

Figure 1 A cobalt (III) aspartic acid complex.

In general, a decrease in reduction potential upon complex formation has been observed. This means that the higher oxidation states of the metal ion become more stable after complexation. This change in reduction potential, although not popular in synthetic coordination chemistry, can be made as a basis for the detection of complex formation.

6.15 Amperometry

Amperometry involves the measurements of currents at constant voltage applied at the dropping mercury electrode. The value of electrode potential is chosen in such a way that only the metal ion is reduced. This method is generally used for the determination of metal ion present in aqueous solution. An aqueous solution of Zn^{+2} may be titrated with an EDTA solution at an applied electrode potential of -1.4 V.

The plot of volume of ligand versus current in this case is shown in Figure 2(a). From the graph, it is seen that there is a decrease in current with an increase in ligand concentration, i.e. decrease in free metal ion concentration, and it finally attains a constant minimum value. This trend in the curve is an indicative of the complex formation.

A V-shaped curve is obtained when an electrode potential is kept where both the metal ion and the ligand are reduced. An amperometric titration of

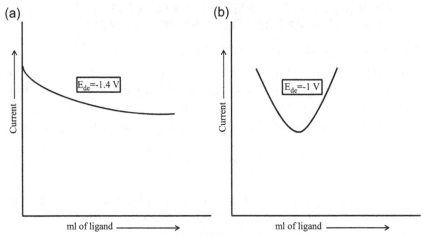

Figure 2 Plots of volume of ligand versus current (a) $Zn^{+2} \rightarrow$ EDTA (b) $Pb^{+2} \rightarrow$ dichromate.

Pb^{+2} with standard dichromate solution at an electrode potential of -1 V shows the graph with a V-shaped curve, as shown in Figure 2(b).

6.16 Electromotive Force (EMF)

Electromotive force (EMF) is the voltage generated by battery, which results into the flow of current. It is measured in volts. EMF of any system depends upon the nature of the ion present in it. Due to the change in the type of ions, an appreciable change in the EMF of the system occurs as a result of complex formation.

6.17 Polarography

The amperometric and polarographic methods rely on the same principle that the diffused current (i_d) is proportional to the concentration. Thus, when an electroactive material (metal ion) is removed from a solution using some reagent (ligand), a decrease in the diffused current is observed. From the plot of current versus voltage, known as polarogram, important information pertaining to the formation of a complex can be obtained. The characteristic half-wave potential of a simple metal ion is shifted when the metal ion undergoes complex formation. The information from the polarography can also be useful in determining the stability constants and coordination number of the complex.

7. EXPERIMENTAL DETERMINATION OF STABILITY CONSTANT AND COMPOSITION OF A COMPLEX

7.1 Spectrophotometric Method [5]

This method is based on Beer's law, which says that $A = \varepsilon \times l \times c$, where A is the optical density or absorbance, ε (epsilon) is the mole extinction coefficient, l is the length of the cell and c is the concentration.

Consider the formation of a complex given by the following reaction:

$$M^{+n} + L \overset{K_f}{\rightleftarrows} ML^{+n} \quad K_f = \frac{[ML^{+n}]}{[M^{+n}][L]}$$

The complex $[ML^{+n}]$ exhibits a characteristic absorption pattern depending on its composition. The wavelength at which the complex absorbs most is termed as λmax. The value of the formation constant K_f can be obtained by measuring the absorbance (A) of the sample solution at the λ_{max} of the complex.

According to the standard procedure, a known quantity of metal ion and ligand are equilibrated. Under equilibrium conditions, both metal and ligand can exist in the solution, either in the free state or complexed form. Thus, the total metal concentration $[C_M]$ and total ligand concentration $[C_L]$ can be given as the following:

$$[C_M] = [M^{+n}] + [ML^{+n}] \text{ and } [C_L] = [L] + [ML^{+n}]$$

or

$$[M^{+n}] = [C_M] - [ML^{+n}] \text{ and } [L] = [C_L] - [ML^{+n}]$$

The term 'c' in the Beer's equation can be replaced by $[ML^{+n}]$ in this case.

Thus, Beer's equation can be shown as $[ML^{+n}] = \frac{A}{\varepsilon \times l}$

Hence,

$$[M^{+n}] = [C_M] - \frac{A}{\varepsilon \times l} \text{ and } [L] = [C_L] - \frac{A}{\varepsilon \times l}$$

The values from these two equations enable the solution of the equation for the formation constant.

7.1 Job's Method

Analogous to the spectrophometric method, the Job's method also involves the measurement of absorbance of the complex at its λ_{max}. This method involves the preparation of several sets with constant volume and variable concentrations of the complexes. Due to this, this method is also known as the method of continuous variation. Using this method, it is possible to determine the metal-ligand ratio (composition) of a complex [1]. The standard procedure involves preparation of 10 sets of solutions of a complex. Different volumes of metal and ligand solutions are equilibrated in these sets in such a way that the total volume of every set remains the same. An illustration of such a set preparation is shown in Table 3:

Table 3 Preparation of sets for a Job's method experiment

Set No.	1	2	3	4	5	6	7	8	9	10
M^{+n} (ml)	0	1	2	3	4	5	6	7	8	9
L (ml)	10	9	8	7	6	5	4	3	2	1
Total (ml)	10	10	10	10	10	10	10	10	10	10

In each of the sets, the sum (C) of total concentration of the metal ion (C_M) and total concentration of the ligand (C_L) remains constant (i.e. $C = C_M + C_L$ remains constant).

The graph of the mole fraction of ligand ($mf_L = C_L/C$) against absorbance (A) measured at the characteristic λ_{max} will appear as shown Figure 3.

The mole fraction of ligand corresponding to the maximum absorbance can be identified by extrapolating the legs of the curve. From Beer's equation, the absorbance is directly proportional to the concentration of the complex. Hence, this point corresponds to the composition with the maximum concentration of the complex.

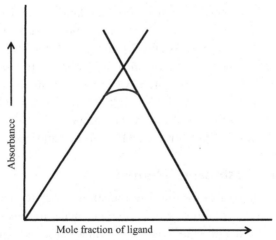

Figure 3 A plot of mole fraction of ligand → absorbance measured at a characteristic λ_{max}.

For a complex ML_n, 'n' represents the number of ligands per metal ion and can be mathematically equated as C_L/C_M. For example, if the complex is represented by a formula ML, the value of 'n' is 1, and it is said that the metal ligand ratio in this case is 1:1. In such a case, the concentration of the complex will be at a maximum in set number six, which contains an equal volume of metal and ligand solutions. The mole fraction of ligand corresponding to this set can give the value of 'n' using the following equation:

$$n = \frac{mf_L}{1 - mf_L}$$

For set number 6, $mf_L = C_L/C = 5/10 = 0.5$
Thus,

$$n = \frac{mf_L}{1 - mf_L} = \frac{0.5}{1 - 0.5} = 1$$

Even though Job's method is a user-friendly method, it is necessary to make sure before using this method that the metal and ligand do not make complexes of more than one composition. Under certain circumstances, it becomes very difficult to create several sets with a constant volume. The application of Job's method is not possible in such cases.

7.2 Mole Ratio Method

This method is a similar, yet simpler, version of the Job's method. The difference between the two methods is that in the molar ratio method, the total analytical concentration of metal/ligand is kept constant instead of the sum of ligand and metal concentration. The absorbance is measured at a wavelength, where the complex absorbs strongly but the ligands and metal ion do not. A plot between absorbance versus ligand to the metal concentration ratio (C_L/C_M) is prepared as shown in Figure 4. Extrapolating the straight line portions until they cross, the value of $R = (C_L/C_M)$ gives the ratio of ligand to the metal ion in the complex.

Hence, a formula of the complex ML_n can be easily determined.

7.3 Matching Absorbance Method

This method is applicable for the determination of stability constants. The absorbance of two solutions with different metal to ligand ratios are measured at a wavelength where only the complex absorbs strongly, but the ligand and metal ion do not. The solution showing higher absorbance is

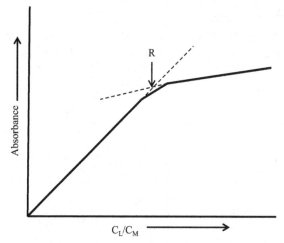

Figure 4 A plot of absorbance → ligand to the metal concentration ratio.

diluted to an extent until the absorbance of the two solutions becomes the same. Let the concentration of metal and ligand in solution (1) be C_{M1} and C_{L1}, respectively, and solution (2) C_{M2} and C_{L2}.

Now, for a 1:1 complex, we have

$$[M_1] = C_{M1} - [ML]_1$$
$$[L_1] = C_{L1} - [ML]_1$$
$$[M_2] = C_{M2} - [ML]_2$$
$$[L_2] = C_{L2} - [ML]_2$$

When the absorbance of the two solutions becomes equal, then $[ML]_1 = [ML]_2 = C_s$, where C_s is some unknown concentration. The stability constant K can be calculated from the stability constant expression, which contains C_s and K as unknown by eliminating C_s. Thus,

$$K = \frac{[ML_1]}{[M_1][L_1]} = \frac{[ML_2]}{[M_2][L_2]}$$

$$= \frac{C_S}{(C_{M1} - C_S)(C_{L1} - C_S)} = \frac{C_S}{(C_{M2} - C_S)(C_{L2} - C_S)}$$

And

$$C_S = \frac{(C_{M1} C_{L1}) - (C_{M2} C_{L2})}{(C_{M1} + C_{L1}) - (C_{M2} + C_{L2})}$$

Thus, the stability constant can be easily determined.

7.4 Bjerrum Method

Bjerrum determined the stability constants of a large number of metal complexes forming in aqueous solutions. He described the stepwise formation of a series of metal complexes of the type ML, $ML_2 \ldots ML_n$ by the following equations:

$$M + L \rightarrow ML \qquad K_1 = \frac{[ML]}{[M][L]}$$

$$M + L \rightarrow ML_2 \qquad K_2 = \frac{[ML_2]}{[M][L]}$$

$$ML_{n-1} + L \rightarrow ML_n \quad K_n = \frac{[ML_n]}{[ML_{n-1}][L]}$$

The equation for the formation constant corresponding to the first step can also be written as

$$[ML] = K_1[M][L]$$

Similarly for the second step,

$$[ML_2] = K_1[ML][L]$$

Substituting the value of [ML] from the first equation into the second, we get

$$[ML_2] = K_1 K_2 [L]^2 [M]$$

Thus,

$$[ML_n] = K_1, K_2, K_3, \ldots K_n [L]^n [M]$$

A function $\bar{n}$ is defined as the average number of ligand molecules bound per mole of metal, and it is expressed as the following:

$$\bar{n} = \frac{[ML] + 2[ML_2] + 3[ML_3] + \ldots + n[ML_n]}{[M] + [ML] + [ML_2] + [ML_3] + \ldots + [ML_n]}$$

Substituting the values of the complex concentrations from the previous equations, we get

$$\bar{n} = \frac{K_1[M][L] + 2K_1 K_2[M][L]^2 + \ldots + nK_1 K_2 K_3 \ldots K_n[M][L]^n}{[M] + K_1[M][L] + K_1 K_2[M][L]^2 + \ldots + K_1 K_2 K_3 \ldots K_n[M][L]^n}$$

$$\therefore \bar{n} = \frac{K_1[L] + 2K_1K_2[L]^2 + \ldots + nK_1K_2K_3\ldots K_n[L]^n}{1 + K_1[L] + K_1K_2[L]^2 + \ldots + K_1K_2K_3\ldots K_n[L]^n}$$

This equation is known as the Bjerrum formation function equation. If the concentration of unbound ligand can be known experimentally, $\bar{n}$ can be calculated from the equation: $\bar{n} = \frac{L_t - [L]}{M_t}$

where L_t and M_t are the total concentration of ligand and metal, respectively. The solution of this equation for known values of n and corresponding (L) values gives the values of the formation constants, i.e. K_1, K_2, $K_3 \ldots K_n$.

An illustration of the Bjerrum method for calculating the formation constants for a complex between Cu^{+2} and 5-sulphosalicyclic acid (H_3L) is as follows.

7.4.1 Procedure

The procedure involves titration of 100 ml of each solution $CuSO_4$ and 5-sulphosalicyclic acid with a standardized solution of sodium hydroxide. A graph of pH against the volume of sodium hydroxide is plotted as shown in Figure 5(a).

Note the point of inflexion of the curve, which corresponds to the mixed solution. Calculation of the horizontal distance between the curves are used to find values of $\bar{n}$. A formation curve of $\bar{n}$ against $-\log [L^{3-}]$,

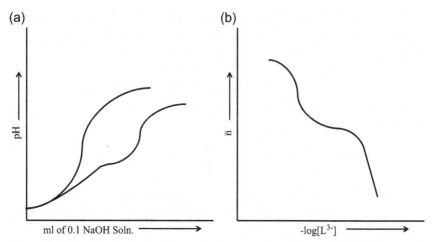

Figure 5 (a) Graph of pH → volume of sodium hydroxide (b) A formation curve of $\bar{n}$ → $-\log [L^{3-}]$.

shown in Figure 5(b), is drawn by obtaining the values of $[L^{3-}]$ found at various pH from the following equation:

$$[L^{3-}] = \frac{[HL_3]_{total} - [CuL^-] - 2[CuL_2^{4-}]}{\frac{[H^+]^2}{[K_2K_3]} + \frac{[H^+]}{[K_3]} + 1}$$

where K_2 and K_3 found from the formation curve of HL_3, which are 3.23×10^{-3} and 1.81×10^{-3}, respectively.

The values of K_1 and K_2 found from the formation curve at $\bar{n} = 0.5$ and $\bar{n} = 1.5$ are $K_1 = 2.2 \times 10^9$ and $K_2 = 6.3 \times 10^6$.

7.5 Polarographic Method

This method is based on an observation that the reduction potentials of metal ions are reduced as a result of complex formation. Consider the reaction:

$$ML_n + xe + Hg \overset{K}{\rightarrow} MHg + nL$$

This reaction, occurring in two steps, can be shown as the following:

$$ML_n \overset{K_1}{\rightarrow} M + nL$$

and

$$M + xe + Hg \overset{K_2}{\rightarrow} MHg$$

Here, K is the overall formation constant, while K_1 and K_2 are the stepwise formation constants corresponding to the respective steps. Also, M is the metal, L the ligand, MLn is the complex and the value of x indicates the number of electrons involved in the reduction.

A relation between the potential at the dropping mercury cathode (E), the half-wave potential ($E_{0.5}$), the current (i) and the diffusion current (i_d) can be given as the following:

$$E = E_{0.5} - \frac{0.0591}{x} \log \frac{i}{i_d - i}$$

A plot of E against ($i/i_d - i$) shows a straight line for a reversible potential. The slope (m) of this line for an equation of this type comes to $0.059/x$, where x is the number of electrons transferred in the reduction.

$$\text{At } 25°C, \quad E_{0.5} = E^0 - \frac{0.0591}{x} \log K - \frac{0.0591}{x} \log \frac{K_1}{K_2} - n \frac{0.0591}{x} \log[L]$$

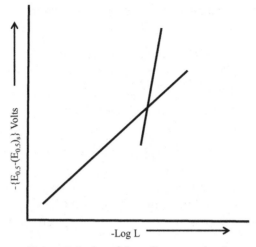

Figure 6 A plot of $E_{0.5} - E_{(0.5)s} \rightarrow -\log L$.

Where E^0 is the standard potential of the metal–amalgam electrode. The value of K_1/K_2 depends on the characteristics of the electrode and conditions during the experiment. n is the number of coordinated ligands and [L] is the concentration of ligands.

The slope of the line obtained in the plot of $E_{0.5}$ versus log [L], comes to $-0.059\, n/x$. The value of n can be calculated, as the value of x is known.

$$E_{0.5} - E_{(0.5)s} = -\frac{0.0591}{x}\log K - n\frac{0.0591}{x}\log[L]$$

Where $E_{(0.5)s}$ is the half-wave potential in the absence of ligands.

The plot of $E_{0.5} - E_{(0.5)s}$ against $-\log L$ yields a straight line as shown in Figure 6, the intercept of which on the coordinate axis gives $(0.0591/x)$ log K, and hence, k can be determined. It is also clear that the slope of the line is $-n0.0591/x$ so that the value of n can be calculated.

7.6 Solubility Method

This method is particularly useful in determining the stability constants of the complexes containing the metal ion capable of forming a sparingly soluble salt whose solubility product is known. Moreover, the complex formed should be fairly soluble. The stability constant of the complex can be determined by measuring the increase in solubility of the salt caused by the presence of the ligand. Consider the reaction of silver acetate with acetate ion. If a sodium or potassium acetate solution of known high concentration is equilibrated at a constant temperature with an excess of silver acetate,

$$AgOAc \rightleftarrows Ag^+ + OAc^-$$

$$K_{sp} = [Ag^+][OAc^-]$$

$$Ag^+ + 2OAc^- \rightleftarrows Ag(OAc)_2^-$$

$$\beta_2 = \frac{[Ag(OAc)_2^-]}{[Ag^+][OAc^-]^2}$$

After the equilibrium is achieved, the excess solid silver acetate is removed by filtration and the solution is analysed for total silver. Then, the total silver concentration in the solution is given as $C_{Ag^+} = [Ag^+] + [Ag(OAc_2)^-]$

From the expression for the solubility product, $[Ag^+] = K_{sp}/C_{NaOAc}$

Thus, the equation for the formation constant becomes

$$\beta_2 = \left(C_{Ag^+} - K_{sp}/C_{NaOAc}\right) / \left[(K_{sp}/C_{NaOAc} C_{NaOAc})^2\right]$$

Since all the remaining quantities are known, the overall stability constant β_2 can be easily calculated.

7.7 Electromotive Force Method

A metal in contact with a solution of ions gives rise to a potential (E) as shown by

$$E = E^0 - (RT/nF)\ln([M^{n+}]/[M]\text{solid})$$

If two such cells with different metals are connected, the difference in voltage of the two cell potentials can be measured. If both the cells are identical (same metal and the same concentration), the measured voltage difference will be equal to zero. Now, if a ligand solution is added to one of the cells due to the complex formation, a net voltage difference will be observed. This observed difference in voltage can be given as the following:

$$E_{obs} = E_1 - E_2 = -(RT/nF) \ln [M^{2+}]_1/[M]_{solid} - \ln([M^{2+}]_2/[M]_{solid})$$
$$= -(RT/nF) \ln([M^{2+}]_1/[M^{2+}]_2)$$

Where T is the absolute temperature, R is the gas constant, F is the Faraday (96,500 coulombs) and n is the number of electrons transferred. Thus, the observed voltage is directly related to the free ion concentrations in the cells. In actual experiments, the fixed volume of ligand is added successively to the metal ion solution in one cell, and the corresponding voltages are measured. The stability constant can be evaluated from the knowledge of total concentration of the metal ion, ligand and uncomplexed metal ion.

8. EXERCISES

8.1 Multiple Choice Questions

1. Which of the following complexes has a fast rate of reaction?
   (a) unstable   (b) stable
   (c) labile   (d) inert

2. Which of the following complexes does not easily react?
   (a) octahedral   (b) labile
   (c) stable   (d) unstable

3. What does $1/K$ values stand for?
   (a) Boltzman constant   (b) variable
   (c) instability constant (d) stability constant

4. The complexes with large activation energy are found to be
   _____.
   (a) reactive   (b) octahedral
   (c) inert   (d) labile

5. Which of the following ligands will form more stable complexes?
   (a) I^- (b) Cl^-
   (c) F^- (d) Br^-

6. The Stability of a complex _____ with an increase in number of chelate rings.
   (a) remains constant (b) vanishes
   (c) increases   (d) decreases

7. Which of the following methods is used to find the composition of a complex?
   (a) Dumas method   (b) Spectrophotometric method
   (c) Job's method   (d) Bjerrum method

8. The Solubility of AgCN _____ due to addition of KCN.
   (a) increases   (b) decreases
   (c) does not change   (d) becomes zero

9. An acidified aqueous solution of zinc salt contains Zn as _____.
   (a) cation   (b) atom
   (c) anion   (d) all of these

10. What will be the value for the dipole moment for a molecule in which all the bond moments are non-zero?
    (a) non-zero   (b) zero
    (c) negative   (d) (a) or (b)

11. The square planar complexes of Ni(II) are _____.
    (a) paramagnetic   (b) diamagnetic
    (c) ferromagnetic   (d) none of these

12. ESR technique is useful in studying complexes containing _____ electrons.
    (a) free   (b) paired
    (c) extra   (d) unpaired

13. The electrophoresis technique is useful to detect the formation of _____ complexes.
    (a) polynuclear   (b) cationic
    (c) neutral   (d) anionic

14. Which type of shape of the curve is observed in an amperopetric plot when the electrode potential is kept at the value where both metal and ligand are reduced?
    (a) 'V' shaped   (b) 'C' shaped
    (c) 'U' shaped   (d) 'L' shaped

15. Which of the following denotes Bjerrum formation function?
    (a) K (b) Δ
    (c) β (d) $\bar{n}$

16. The reduction potential _____ as a consequence of complex formation.
    (a) fluctuates   (b) increases
    (c) remains constant   (d) decreases

8.2 Short/Long Answer Questions

1. Compare and contrast the thermodynamic and kinetic stability.
2. Explain inert and labile inner orbital octahedral complexes according to VBT.
3. Explain inert and labile octahedral complexes according to CFT.

4. Mention factors affecting the lability of complexes in addition to the electronic configuration of the central metal ion forming the complexes, giving one example of each.
5. Explain the effect of geometry on the lability of complexes.
6. Mention all the factors affecting the stability of complexes.
7. Explain the effect of steric factors of ligand on the stability of complexes.

SUGGESTED FURTHER READINGS

The topics discussed in this chapter are a part of a standard graduate curriculum. A majority of the textbooks with titles related to coordination chemistry, inorganic chemistry and general chemistry can act as a source of further reading. In order to satisfy the thirst of further information on the bases of the detection of complex formation, the reader must read specialized books on the techniques mentioned in the text. There are several web resources useful for further learning. Some of them are listed below.

http://nptel.ac.in/courses/104106063/Module%205/Lectures%208-10/Lectures%208-10.pdf

http://classes.uleth.ca/200403/chem3820a/3820%20lecture%20chapter_9_part1_2004.pdf

http://www.ncbi.nlm.nih.gov/pmc/articles/PMC4028694/

http://www.nptel.ac.in/courses/104106064/

REFERENCES

The following articles published in journals contain advanced applications of the concepts discussed in the present chapter:

[1] Saxena MC, Bhattacharya AK. Complex formation between trivalent cerium and alloxan. Z für Anorg Allg Chem 1962;315(1–2):114–7.

[2] Rezayi M, et al. Conductance studies on complex formation between c-methylcalix[4] resorcinarene and titanium(III) in acetonitrile-H_2O binary solutions. Molecules 2013; 18(10):12041.

[3] Victor AH. Separation of nickel from other elements by cation-exchange chromatography in dimethylglyoxime/hydrochloric acid/acetone media. Anal Chim Acta 1986;183:155–61.

[4] Strelow FWE. Distribution coefficients and cation-exchange behaviour of some ammines and aquo complexes of metallic elements in ammonium nitrate solution. Anal Chim Acta 1990;233:129–34.

[5] Khan AAP, et al. Spectrophotometric methods for the determination of ampicillin by potassium permanganate and 1-chloro-2,4-dinitrobenzene in pharmaceutical preparations. Arabian J Chem 2015;8(2):255–63.

CHAPTER 5

Reactions in Octahedral Complexes

Contents

1. INTRODUCTION

A reaction is a phenomenon which follows the action. In the present studies, the scope of study is the action between metal and ligand to form a complex. In order to perform a reaction, the action must be studied and

Essentials of Coordination Chemistry
http://dx.doi.org/10.1016/B978-0-12-803895-6.00005-7

understood well. The ΔG^0 and ΔH_f^0 corresponding to the formation actions provide an insight about the extent of energy required to carry out reactions. Substitutions, redox and isomerization are the main types of reactions studied in the octahedral complexes. The majority of these reactions can easily be classified as substitution reactions. The metals in complexes are Lewis acids, while the ligands are Lewis bases. Hence, the ligand substitution reactions are nucleophilic substitution reactions, whereas the metal substitution reactions are the electrophilic substitution reactions. The metal in a complex is attached to all the ligands, which requires the breaking of all the metal—ligand (M—L) bonds for a metal substitution. In such cases, the reaction is rather decomposition and the formation of yet another complex. Due to this, only the ligand substitution reactions have drawn much attention of coordination chemists. In the following section, various mechanisms for substitution reactions, redox reactions and isomerization reactions are discussed.

2. POTENTIAL ENERGY CURVES SHOWING ENERGY CHANGES INVOLVED IN ENDOTHERMIC AND EXOTHERMIC REACTIONS

Let us consider a reaction. $A + B - C \rightarrow A - B + C$

In this reaction, the species A and B—C are reactants, while A—B and C are the products. The potential energy changes involved during this transformation are presented graphically in two figures. From Figure 1, it can be seen that the reactants are converted into the products through a transition state, which involves the formation of an activated complex. This activated complex is a high-energy molecule with partial bonds. Thus, it is not a true molecule and cannot be isolated due to extremely low stability. The energy required for the formation of this activated complex is the difference between the energy of the reactants and the energy of the activated complex. This energy is known as activation energy. From Figure 1, it is also seen that, apart from the reaction being endothermic or exothermic, the reactant molecules must acquire activation energy for successful transformation. For exothermic reaction, the energy of the product is lower than the energy of the reactants. This reaction energy (ΔH) is the energy released at the successful transformation. Thus, in the case of exothermic reactions, if a few reactant molecules are initially activated externally, the energy released due to the transformation to products

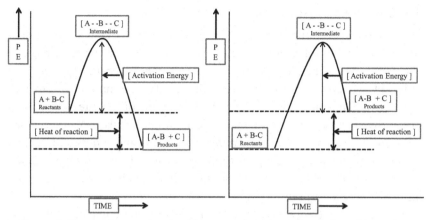

Figure 1 Energy changes involved in an exothermic and endothermic reaction.

becomes the driving force for the excitation of some more reactant molecules, and the reaction continues without an external supply of energy. While in the case of endothermic reactions, with the energy of the products being higher than the reactants, the reactants must be continuously supplied external energy for the reaction to continue.

A reactant molecule, in which some bonds are broken and some new are formed, is known as substrate. Thus, in the mentioned reaction, $B-C$ is the substrate. This reaction proceeds when 'A' attacks on the substrate, hence the species 'A' in the reaction is called the attacking reagent. If the attacking reagent is electron-loving (electrophile), such as nitrosonium ion (NO^+) or bromonium ion (Br^+), it is deficient of electrons and hence will attempt an electrophilic substitution reaction on the substrate. More commonly in coordination chemistry, the attacking reagents are nucleus-loving (nucleophiles), such as chloro ligand (Cl^-), hydroxo ligand (OH^-) or cyano ligand (CN^-), which are electron-rich and tend to replace some other ligand on the substrate molecule by nucleophilic substitution reaction.

3. MECHANISMS OF NUCLEOPHILIC SUBSTITUTION REACTIONS IN OCTAHEDRAL COMPLEXES

Nucleophilic substitution (SN) reactions in octahedral complexes are of paramount importance in coordination chemistry. These reactions follow any of the two popular mechanisms: SN^1 and SN^2.

3.1 SN¹ Mechanism

In the abbreviation, the letter 'S' indicates that the reaction is a substitution, and the letter 'N' indicates the nucleophilic nature of the attacking reagent. The letter '1' is often misunderstood and is considered to represent first order. However, '1' means that the molecularity (the number of different types of reactant molecules involved in a reaction) of the rate determining step (slowest step) in this mechanism is one (unimolecular). A nucleophilic substitution reaction involves the removal of one ligand and its replacement by another ligand. According to this mechanism, shown in Figure 2, the first step involves the removal of outgoing ligand. Therefore, this mechanism is also called the dissociation mechanism.

Consider a nucleophilic substitution reaction in a complex with coordination number six (CN = 6) octahedral (Oh) substrate MX_5Y occurring via SN^1 mechanism.

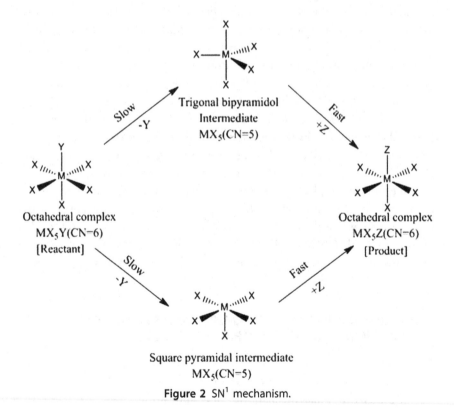

Figure 2 SN¹ mechanism.

$$\underset{\text{Oh,CN=6}}{MX_5Y} + Z \xrightarrow{SN^1} \underset{\text{Oh,CN=6}}{MX_5Z} + Y$$

According to SN^1 mechanism, the first step involves the loss of the leaving group Y. This breaking of the M$-$Y bond results into formation of a five-coordinated (trigonal bipyramidal/square pyramidal) intermediate. This step is slow and hence a rate determining step.

$$\underset{\text{Oh,CN=6}}{MX_5Y} \xrightarrow[\text{molecularity=1}]{\text{slow(R.D.step)}} \underset{\text{tbp/sq.py,CN=5}}{MX_5} + Y$$

It should also be noted that the molecularity of this step is one.

At the onset of the second step, the vacancy created on the substrate due to the loss of Y facilitates the entry of the incoming ligand Z. Thus, in a rapid bimolecular step, the product MX_5Z is obtained.

$$\underset{\text{tbp/sq.py,CN=5}}{MX_5} + Z \xrightarrow[\text{molecularity=2}]{\text{fast}} \underset{\text{Oh,CN=6}}{MX_5Z}$$

Since the rate of the slowest step determines the overall rate of the reaction, and the rate of the slowest step (step one) depends only on the concentration of the substrate MX_5Y.

The overall rate of reaction $= K \cdot [MX_5Y]$

3.2 SN^2 Mechanism

In the abbreviation, '2' means that the molecularity of the rate determining step (R.D. step) in this mechanism is two (bimolecular). According to this mechanism, the first step involves the entry of the attacking nucleophile Z. Therefore, this mechanism is also called association mechanism.

Consider a nucleophilic substitution reaction in a six-coordinated (CN = 6) octahedral (Oh) substrate MX_5Y, occurring via SN^2 mechanism.

$$\underset{\text{Oh,CN=6}}{MX_5Y} + Z \xrightarrow{SN^2} \underset{\text{Oh,CN=6}}{MX_5Z} + Y$$

According to SN^2 mechanism, the first step involves the addition of the incoming ligand Z. The making of this additional M$-$Z bond results into the formation of a seven-coordinated (pentagonal bipyramidal) intermediate. This step is energy expensive, sterically unfavoured and hence a slow (R.D.) step.

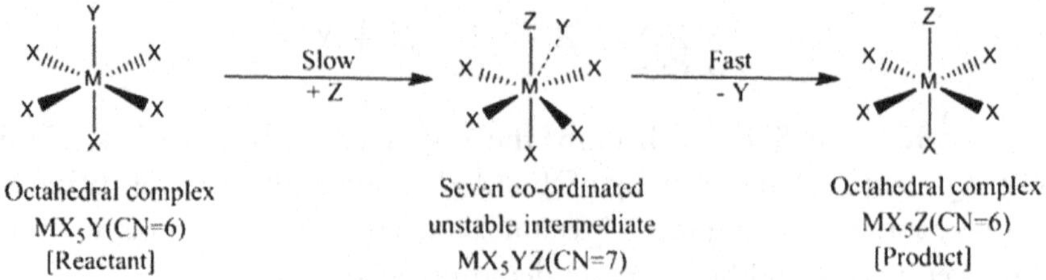

Figure 3 SN^2 mechanism.

$$MX_5Y + Z \xrightarrow[\substack{\text{molecularity}=2}]{\text{slow(R.D.step)}} MX_5YZ$$

$\underset{\text{Oh,CN=6}}{MX_5Y} + Z \qquad \underset{\text{Pentagonal bipyramidal,CN=5}}{MX_5YZ}$

It should also be noted that the molecularity of this step is two.

At the onset of the second step, the crowding created on the substrate due to an additional position occupied by Z facilitates the removal of the leaving ligand Y. Thus, in a rapid unimolecular step, the product MX_5Z is obtained.

$$\underset{\substack{\text{Pentagonal bipyramidal,CN=5}}}{MX_5YZ} \xrightarrow[\text{molecularity}=1]{\text{fast}} \underset{\text{Oh.,CN=6}}{MX_5Z + Y}$$

Since the rate of the slowest step determines the overall rate of the reaction and the rate of the slowest step (step one) depends both on the concentrations of the substrate MX_5Y and the nucleophile Z.

The overall rate of reaction $= K \cdot [MX_5Y] \cdot [Z]$

Comparing the equations for the rate of reactions for SN^1 and SN^2 mechanisms, it is apparent that if the rate depends only on the concentration of the substrate, the mechanism followed for a particular reaction is SN^1. While, if the rate depends on the concentrations of the substrate and the attacking reagent, the probable mechanism is SN^2.

A schematic diagram for an SN^2 mechanism is shown in Figure 3.

4. MECHANISMS INVOLVED IN REACTIONS LIKE ACID HYDROLYSIS AND BASE HYDROLYSIS OF SIX-COORDINATED CO (III) AMMINE COMPLEXES

4.1 Acid Hydrolysis and Base Hydrolysis

The water molecule can coordinate with metal ions in two different forms, viz, OH_2 and OH^-. The substitution reaction, in which the water molecule replaces any other ligand, is called hydrolysis. The replacement of a ligand by a water molecule as OH^- yields a hydroxo complex, and the reaction is termed as base hydrolysis. The base hydrolysis occurs at a pH above 10.

$$[Co(NH_3)_5Cl]^{2+} + OH^- \xrightarrow{\text{Base hydrolysis}} [Co(NH_3)_5OH]^{2+} + Cl^-$$

The replacement of a ligand by a water molecule as H_2O yields an aquo complex, and the reaction is termed as acid hydrolysis or aquation. The acid hydrolysis occurs at a pH below 3.

$$[Co(NH_3)_5Cl]^{2+} + H_2O \xrightarrow{\text{Acid hydrolysis}} [Co(NH_3)_5(H_2O)]^{3+}$$

4.2 Mechanism of Acid Hydrolysis of Six-Coordinated Co (III) Ammine Complexes

In this reaction carried out in an aqueous medium, water plays a double role, viz, solvent and nucleophile. The species $[Co(NH_3)_5Cl]^{2+}$ is of particular interest because the ammonia ligand is not readily replaced by water molecules, which leaves the possibility of replacement of only the chloro ligand. From the rate laws corresponding to the SN^1 and SN^2 mechanisms, we have seen that it is the effect of change in the nucleophile concentration on the rate that allows us to know the mechanism.

While carrying out this reaction in an aqueous medium, it is impossible to change the concentration of water (nucleophile). The concentration of water in aqueous solutions is always constant and is 55.5M. The inability to change the nucleophile concentration does not permit us to distinguish the rate laws for SN^1 and SN^2 mechanisms. Thus, rate laws cannot provide any information regarding the mechanism of acid hydrolysis carried out in an aqueous medium.

4.3 Mechanism of Base Hydrolysis of Six-Coordinated Co (III) Ammine Complexes

Consider a base hydrolysis given by the reaction

$$[Co(NH_3)_5Cl]^{2+} + OH^- \xrightarrow{\text{Base hydrolysis}} [Co(NH_3)_5OH]^{2+} + Cl^-$$

If this reaction follows an SN^2 mechanism,

$$[Co(NH_3)_5Cl]^{2+} + OH^- \xrightarrow[\text{slow,RDstep}]{\text{step1}} [Co(NH_3)_5(OH)Cl^-]^+$$

$$\xrightarrow[\text{fast}]{\text{step2}} [Co(NH_3)_5(OH)]^{2+} + Cl^-$$

The rate of reaction = K[substrate][nucleophile] = $K[[Co(NH_3)_5Cl^{+2}]]$ $[OH^-]$.

The other probable mechanism for this reaction is SN^1CB mechanism. According to this mechanism, the ammine ligand of the substrate gives up a proton to form a conjugate base (CB) before it takes up the SN^1 pathway, i.e.

$$NH_3 \rightarrow NH_2^- + H^+$$

Thus,

$$\left[Co^{+3}(NH_3)_5Cl\right]^{+2} + OH^- \underset{}{\overset{fast}{\rightleftharpoons}} \frac{\left[Co^{+3}(NH_3)_4(NH_2)Cl^-\right]^+}{Conjugate base\ (CB)} + H_2O$$

The equilibrium constant for the above reaction is given as

$$K = \frac{[CB][H_2O]}{\left[Co^{+3}(NH_3)_5Cl\right]^{+2}[OH^-]}$$

$$\therefore [CB] = \frac{K\left[Co^{+3}(NH_3)_5Cl\right]^{+2}[OH^-]}{[H_2O]}$$

The lability of the conjugate base is higher than that of the original substrate complex. Hence, The CB can now undergo the SN^1 mechanism as follows

$$\frac{\left[Co^{+3}(NH_3)_4(NH_2^-)Cl^-\right]^+}{Conjugate\ base\ (CB)} \xrightarrow[slow,RD]{step\ 1} \left[Co^{+3}(NH_3)_4(NH_2)\right]^{+2} + Cl^-$$

In the second and last step, the five-coordinated intermediate can now quickly react with the water molecules of the aqueous solution to give the product.

$$\left[Co^{+3}(NH_3)_4(NH_2)\right]^{+2} + H_2O \xrightarrow[fast]{step\ 2} \frac{\left[Co^{+3}(NH_3)_5(OH)\right]^{+2}}{hydroxocomplex\ (product)}$$

We know that the slowest step is the rate-determining step. Hence, the overall rate of base hydrolysis should be given by the rate of step one, which is shown below.

$$\text{Rate} = \text{K1}[\text{CB}]$$

$$\text{but } [\text{CB}] = \frac{\text{K}\left[\text{Co}^{+3}(\text{NH}_3)_5\text{Cl}\right]^{+2}[\text{OH}^-]}{[\text{H}_2\text{O}]}$$

$$\therefore \text{Rate} = \frac{\text{K1K}\left[\text{Co}^{+3}(\text{NH}_3)_5\text{Cl}\right]^{+2}[\text{OH}^-]}{[\text{H}_2\text{O}]}$$

Also, $\dfrac{\text{K1K}}{[\text{H}_2\text{O}]} = \text{Kb}$ (Base dissociation const. for the conjugate base)

$$\text{Hence, Rate} = \text{Kb}\left[\text{Co}^{+3}(\text{NH}_3)_5\text{Cl}\right]^{+2}[\text{OH}^-]$$

It is important to note here that even though the SN^1CB mechanism is much like that of SN^1, the rate of reaction depends on the concentrations of both the substrate and nucleophile and is consistent with SN^2 mechanism.

In the Co (III) complexes with ligands containing replaceable proton, i.e. NH_3, en etc., the rate of base hydrolysis is often about a million times faster than acid hydrolysis. In order to establish the mechanism of base hydrolysis in such complexes, several experiments have been carried out. The base hydrolysis of Co (III) complexes with ligands without replaceable proton, such as $[\text{Co(CN)}_5\text{Br}]^{-3}$, was found to be very low and independent of the concentration of the nucleophile. Since the non-leaving ligand does not contain a replaceable proton, the SN^1CB mechanism is not possible. The failure of such complexes to undergo the base hydrolysis is an important indication supporting the operation on SN^1CB mechanism in base hydrolysis. Yet another evidence in favour of SN^1CB mechanism is seen while performing base hydrolysis in aprotic nonaqueous solvents such as dimethyl sulfoxide.

It has been observed that the addition of hydrogen peroxide to a reaction mixture containing $[\text{Co(NH}_3)_5\text{Cl}]^{+2}$ ion and hydroxide ion reduces the rate of the base hydrolysis. Hydrogen peroxide dissociates as

$$H_2O_2 + OH^- \rightarrow H_2O + O_2H^-$$

If the SN^2 mechanism was to operate, the species O_2H^- being a stronger nucleophile than hydroxide ion should give a peroxo complex $[\text{Co(NH}_3)_5(\text{O}_2\text{H})]^{+2}$ as the product. However, a peroxo complex is not formed under such conditions, which rules out the prevalence of the SN^2 mechanism.

For an SN^1CB mechanism, the rate of reaction is directly proportion to the concentration of the nucleophile (OH^-), and the addition of peroxide reduces the hydroxide concentration as shown in the above equation. Experimentally, the addition of hydrogen peroxide reduces the rate of base hydrolysis. Thus, all the above evidences lead to conclude that the base hydrolysis of cobalt (III) ammine complexes occur via SN^1CB mechanism.

5. ANATION REACTION

The reaction in which an anionic ligand replaces a coordinated aquo ligand from a complex is known as an anation reaction.

This is a reverse reaction of acid hydrolysis and may be represented as

$$[MX_5(H_2O)]^n + Y^- \rightarrow [MX_5Y]^{n-1} + H_2O$$

An example of anation reaction is

$$[Co(NH_3)_5(H_2O)]^{+3} + Cl^- \xrightarrow{\text{Anation}} [Co(NH_3)_5Cl]^{+2} + H_2O$$

The rate of this reaction depends on the concentration of the complex, as well as a nucleophile. However, the mechanism of this reaction is not yet established.

6. SUBSTITUTION REACTIONS WITHOUT M–L BOND CLEAVAGE

The decarboxylation of a carbonato complex to give an aquo complex is a classic example of this kind of reaction.

Consider the reaction

$$[Co(NH_3)_5(CO_3)]^+ + 3H_3O^+ \rightarrow [Co(NH_3)_5(H_2O)]^{+3}$$

Primarily, this is an acid hydrolysis. However, the isotopic labelling of oxygen of a nucleophile by (O^{18}) revealed that the oxygen atom from the nucleophile does not enter into the product. The most probable mechanism for this reaction is shown in Figure 4.

The mechanism shows that upon the attack of the proton on the oxygen atom bonded to the metal ion, carbon dioxide is eliminated, accompanied by the release of a water molecule. This forms a hydroxo complex, which quickly takes up a proton from the medium to give the resultant aquo complex. The absence of isotopic oxygen (O^{18}) can be easily justified by the above mechanism.

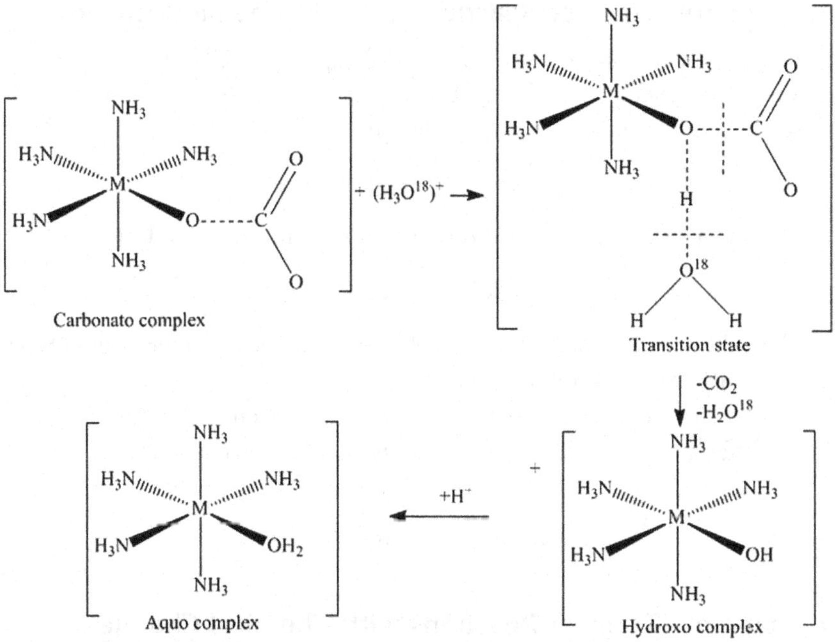

Figure 4 Decarboxylation of a carbonato complex without the cleavage of a metal–ligand bond.

The anation reaction of a cobalt(III) ammine aquo complex using (NO_2^-) as an anionic ligand also occurs without the cleavage of metal–ligand bond.

$$[(NH_3)_5Co-O^{18}H_2]^{+3} + NO_2^- \xrightarrow{\text{Anation}} [(NH_3)_5Co-O^{18}NO]^{+2} + H_2O$$

The retainment of the O^{18} isotope on the aquo ligand in the product complex confirms that the $M-(O^{18})$ bond is not broken.

7. ELECTRON TRANSFER REACTIONS

The electron transfer reactions, also known as oxidation–reduction reactions or redox reactions, involve a transfer of electrons from one atom to the other atom.

The electron transfer reactions may be classified in to two types:

1. Electron transfer reactions without chemical change
2. Electron transfer reactions with chemical change

7.1 Electron Transfer Reactions without Chemical Change

Consider the complex ion pair $[Fe(CN)_6]^{-3}-[Fe(CN)_6]^{-4}$. A process involving the electron exchange between Fe^{+2} and Fe^{+3} with one of the ions labelled by an isotope can be shown as

$$Fe^{+2} + Fe^{*+3} \rightarrow Fe^{+3} + Fe^{*+2}$$

A similar reaction is observed between cerium ions in solution

$$Ce^{+3} + Ce^{*+4} \rightarrow Ce^{+4} + Ce^{*+3}$$

The above reaction does not involve any chemical change. These types of reactions are also known as self-exchange reactions.

An electron exchange reaction occurring between the cerium and iron ions (different ions) shown below is termed as electron transfer cross-reaction.

$$Ce^{+4} + Fe^{+2} \rightarrow Fe^{+3} + Ce^{+3}$$

7.2 Electron Transfer Reactions with Chemical Change

Consider an oxidation–reduction reaction,

$$[Co^{+3}(NH_3)_5Cl]^{+2} + [Co^{+2}(H_2O)_6]^{+2} \rightarrow [Co^{+2}(NH_3)_5Cl]^{+} + [Co^{+3}(H_2O)_6]^{+3}$$

In this reaction, the Co (III) ion bearing ammine and chloro ligands in the reactants binds six aquo molecules in the products. In the same way, Co (II) bearing six aquo ligands binds ammine and chloro ligands in the product. Thus, there is a net chemical change.

An electron transfer cross-reaction of this type can be represented as

$$[Co^{+3}(NH_3)_5Cl]^{+2} + [Cr^{+2}(H_2O)_6]^{+2} + 5H_3O^{+} \rightarrow [Cr^{+3}(H_2O)_5Cl]^{+} + [Co^{+2}(H_2O)_6]^{+3} + 5NH_4^{+}$$

7.3 Mechanism of Electron Transfer Reactions

There are two popular mechanisms for one-electron exchange reactions occurring in octahedral complexes:

1. Inner sphere mechanism, also known as bonded electron transfer or bridge mechanism
2. Outer sphere mechanism, also known as direct electron transfer mechanism

7.4 Inner Sphere Mechanism, also known as Bonded Electron Transfer or Bridge Mechanism

According to this mechanism, the electron transfer and the ligand transfer in the reaction occur at the same time. A close contact between the oxidant and the reductant of the reaction is required. The formation of a ligand bridge affords this intimate contact. The typical bridging ligands involved in such reactions are capable of conveying the electrons and contain more than one lone pair of electrons.

Henry Taube and co-workers discovered the operation of an inner sphere mechanism in a redox reaction involving the oxidation of $[Cr(H_2O)_6]^{+2}$ using $[Co(NH_3)_5Cl]^{+3}$ in an acidic medium (1M $HClO_4$). The reaction is as follows:

$$[Co^{+3}(NH_3)_5Cl]^{+2} + [Cr^{+2}(H_2O)_6]^{+2} + 5H_3O^+ \rightarrow [Cr^{+3}(H_2O)_5Cl]^+$$
$$+ [Co^{+2}(H_2O)_6]^{+3} + 5NH_4^+$$

This reaction involves the following steps:

Step 1: The Cr (III) complex loses a water molecule and links with the Co (II) complex by a chloro bridge, as shown in Figure 5.

The Co (II) complex is the only source of a chlorine atom required for the formation of this bridged activated complex. This has been verified by isotopic labelling. When this reaction is carried out in a medium containing free Cl^- ions and using the Co (II) complex containing isotopically labelled chlorine (^{36}Cl), only labelled chlorine is found in the product.

$$[Co^{+3}(NH_3)_5 {}^{36}Cl]^{+2} + [Cr^{+2}(H_2O)_6]^{+2} + 5H_3O^+ + Cl^-$$
$$\rightarrow [Cr^{+3}(H_2O)_5 {}^{36}Cl]^+ + [Co^{+2}(H_2O)_6]^{+3} + 5NH_4^+$$

Thus, according to the inner sphere mechanism operating in the above reaction, the chloro bridge formed between the two complexes provides a pathway for the transfer of electrons between the metals.

In this example, the formation of the chloride bridge was not observed experimentally but was inferred from the analysis of the product analysis. The Creutz–Taube complex, shown in Figure 6, is a model for the bridged intermediate. This ion was named after Carol Creutz, who prepared it while working with Taube. Pyrazine is the bridging ligand here. In the Creutz–Taube ion, the average oxidation state of Ru is +2.5. Spectroscopic studies have shown that the two Ruthenium centres are equivalent, which indicates the ease with which the electron hole communicates between the two metals. Many more complex mixed valence species are known both as molecules and polymeric materials.

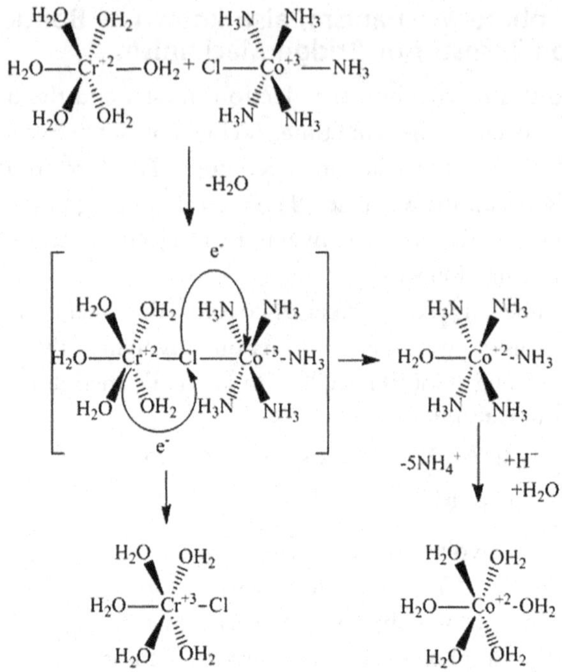

Figure 5 Mechanism of an inner sphere electron transfer reaction.

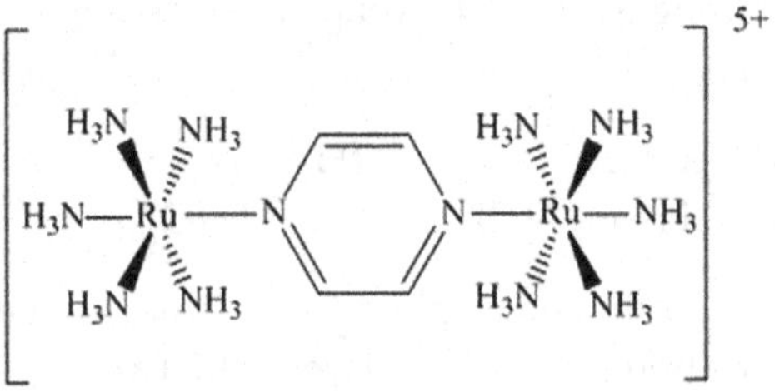

Figure 6 Creutz–Taube complex ion.

Step 2: The electron transfer from Cr (II) to Co (III) in the bridged activated complex occurs through the chloro bridge. This results into oxidation of Cr (II) to Cr (III) and reduction of Co (III) to Co (II).

Step 3: The Cr (III) attracts the Cl$^-$ ion more strongly, as compared to Co (II). Due to this, the chloro ligand becomes a part of the chromium complex in the final product.

Thus, we see that in this reaction Co (III) loses a chloro ligand but gains an electron while Cr (II) gains a chloro ligand but loses an electron.

Experimentally, this reaction is found to be of the first order with respect to both the reactants. Hence,

$$\text{rate} = K[[Co^{+3}(NH_3)_5Cl]^{+2}][[Cr^{+2}(H_2O)_6]^{+2}]$$

It is also seen that the rate of this reaction depends much on the ability of the bridging ligand to provide the path for the electron transfer. It has been observed that the rates of electron transfer reactions with the complexes of the general formula $[Co^{+3}(NH_3)_5X]^{+2}$ vary as $SO_4^{-2} < Cl^- < Br^-$.

A variety of such reactions differing with respect to the number of electrons transferred and number of bridges formed has been studied.

A reaction, $[Pt^{+2}Cl_4]^{-2} + [Pt^{+4}Cl_6]^{-4} \rightarrow [Pt^{+4}Cl_6]^{-4} + [Pt^{+2}Cl_4]^{-2}$, represents an example of a two-electron transfer reaction occurring through a single bridge.

7.5 Outer Sphere Mechanism, also known as Direct Electron Transfer Mechanism

A redox reaction following outer sphere mechanism involves the tunnelling of electrons between the two reactants without creating a significant change in their environments.

In order to understand the outer sphere mechanism, consider a self-exchange reaction

$$[Fe^{+2}(H_2O)_6]^{+2} + [Fe^{+3}(H_2O)_6]^{+3} \rightarrow [Fe^{+3}H_2O)_6]^{+3} + [Fe^{+2}(H_2O)_6]^{+2}$$

From the isotopic labelling and nuclear magnetic resonance (NMR) studies, it has been observed that the reaction has a rate constant consistent with the second-order reaction and occurs rapidly at 25 °C. Since the reaction is without chemical change, there is no change in heat or reactants.

The mechanism involves the intimacy of the two complexes, resulting into the formation of a weak outer sphere complex, as shown in Figure 7. A sufficient overlapping between the acceptor and the donor orbitals of the reactants facilitates the electron transfer.

According to the Frank–Condon principle, the electronic transitions, being very fast, occur without a significant change in the atomic environment.

A plot of nuclear coordinates of the reactants and potential energy shown in Figure 8 indicates that an instantaneous electron transfer is possible when the reactants are at their energy minimum state. The point of

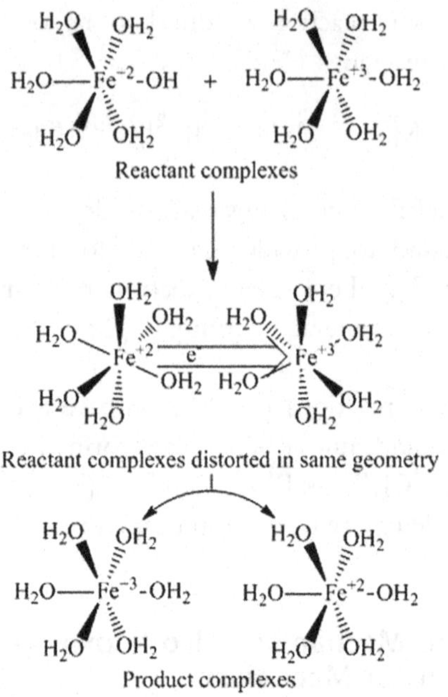

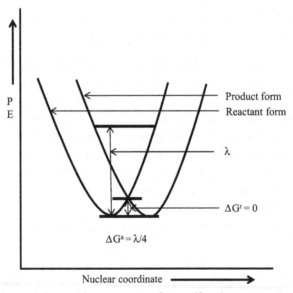

Figure 7 Mechanism of an outer sphere reaction in Fe^{+2}/Fe^{+3} couple.

Figure 8 Potential energy curve for a self-exchange reaction.

intersection of the curve shows the time when the reactants are in the same nuclear configuration and a quick electron transfer occurs. The energy required to reach this position is known as Gibb's activation energy (ΔG^a). Thus, the larger the difference in nuclear configuration of the reactants, the ΔG^a will be more and the transfer of electrons will be slower.

Rudolph Marcus developed the main theory describing the rates of outer-sphere electron transfer in the 1950s. According to this theory, the rate of electron transfer depends on the difference in the redox potentials of the electron-exchanging sites. For most reactions, the rates increases with increased driving force. Another aspect is that the rate depends inversely on the reorganizational energy (λ), which describes the changes in bond lengths and angles that are required for the oxidant and reductant to switch their oxidation states.

According to the Marcus equation

$$k_{ET} = \nu_N K_e e^{-\Delta G^a/RT}$$

Where k_{ET} is the rate of electron transfer and ΔG^a is given by the following equation

$$\Delta G^a = 1/4\lambda \left(1 + \frac{\Delta G^r}{\lambda}\right)^2$$

In these two equations,

k_{ET} is the rate constant for the electron transfer,

ν_N is the nuclear frequency(the frequency at which the reactants encounter in the solution to achieve a transition state),

K_e is the electronic factor (ranges from 0 to 1 and indicates the probability of electron transfer after reaching the transition state),

ΔG^a is the Gibb's activation energy,

R is the gas constant,

T is the temperature of reaction in Kelvin scale,

λ is reorganization energy and

ΔG^r is the Gibb's reaction energy (obtained from the standard potentials of the redox partners, large negative values indicate that the reactions are thermodynamically favourable).

The labels a, b and c, shown in Figure 9 represent the variation of ΔG^a with ΔG^r in a self-exchange reaction ($\Delta G^r = 0$), an activation-less reaction ($\Delta G^r = -\lambda$) and a reaction in which ΔG^a increases as ΔG^r becomes more negative ($\Delta G^r < -\lambda$) respectively.

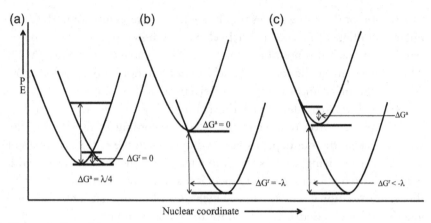

Figure 9 Potential energy curve for different reactions following an outer sphere mechanism.

For a self-exchange reaction (Figure 9(a)), the value of ΔG^r is zero. From the Marcus equation, in such cases, the value of $\Delta G^a = {}^1\!/_4\lambda$. Thus, the reorganization energy λ governs the rate constants of such reactions.

For the redox reactions occurring between different species, ΔG^r remains non–zero (Figure 9(b) and (c)). In the Marcus equation, ΔG^a becomes zero when ΔG^r equals $-\lambda$. Under such conditions, the reaction becomes activation-less due to the mutual cancellation of activation energy and reorganization energy. With larger negative values of ΔG^r, as shown in figure c, the activation energy (ΔG^a) increases according to the Marcus equation, and the rate of the redox reaction decreases. This slowing of the reaction is known as inverted behaviour.

8. ISOMERIZATION REACTIONS

In the presence of a chelating ligand, an isomerization can take place due to the breaking of a metal–ligand bond. An isomerization reaction occurring in tris(3-acetylpentane-2,4-dione)cobalt (III) complex is shown in Figure 10. In order to track the mechanism of the reaction, a set of deuterated hydrogen is placed at the outer methyl group of the complex, as shown in Figure 10. The exchange of the outer CD_3 group with the inner CH_3 group affords the isomerization without the loss of any ligand. It has

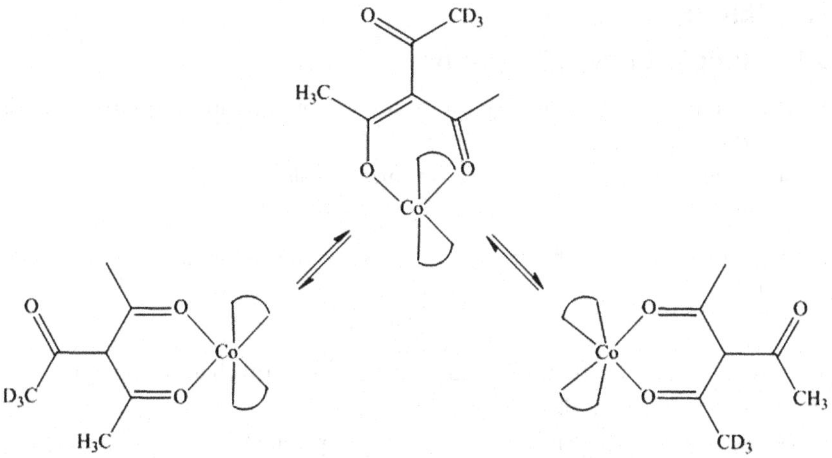

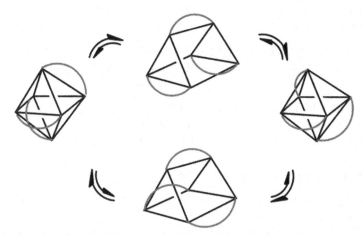

Figure 10 Isomerization reaction in tris(3-acetylpentane-2,4-dione)cobalt(III) complex.

Figure 11 Ray-Dutt twist (top) and Bailar twist (bottom) in the isomerization of an octahedral complex.

been observed that the racemization of $[Ni(en)_3]^{+2}$ takes place by such intramolecular twist.

The two popular mechanisms for the intramolecular rearrangement in octahedral complexes, known as Bailar twist and Ray-Dutt twist occurring through prismatic intermediates, are shown in Figure 11.

9. EXERCISES

9.1 Multiple Choice Questions

1. A reactant in which some of the bonds are broken and some new bonds are formed is called _____.
   (a) electrophile   (b) nucleophile
   (c) substrate   (d) attacking reagent

2. The concentration of water in aqueous solution is nearly _____ M.
   (a) 18   (b) 0.1
   (c) 55.5   (d) 1

3. The geometry of the intermediate during SN^1 mechanism operating on an octahedral complex is _____.
   (a) trigonal bipyramidal   (b) square pyramidal
   (c) pentagonal bipyramidal   (d) (a) or (b)

4. The geometry of the intermediate during SN^2 mechanism operating on an octahedral complex is _____.
   (a) trigonal bipyramidal   (b) square pyramidal
   (c) pentagonal bipyramidal   (d) (a) or (b)

5. An anation reaction is a reverse reaction of _____ reaction.
   (a) hydrolysis   (b) carboxylation
   (c) ammonolysis   (d) addition

6. The base hydrolysis of $[Co(CN)_5Br]^{-3}$ is _____.
   (a) very slow   (b) very fast
   (c) medium   (d) unknown

9.2 Short/Long Answer Questions

1. Compare and contrast the mechanism of SN^1 and SN^2 reactions in octahedral complexes.
2. Why do rate laws fail to give information about the mechanism of acid hydrolysis in octahedral complexes?
3. Giving suitable examples, describe the reactions without M—L bond breaking.
4. Describe the SN^1CB mechanism involved in the base hydrolysis of Co (III) ammine complexes.

SUGGESTED FURTHER READINGS

Due to the classical nature of the content in this chapter, the list of research articles and books that can act as further source of reading is not provided. However, an interested reader can refer to any book dealing with inorganic and coordination chemistry; the topics included in this chapter will invariably be present. A good amount of web resources are available in this area. This resources are more or less the same, but to cater to the taste of different readers, a list of such websites is provided here.

http://www.chem.ox.ac.uk/icl/dermot/mechanism1/lecture2/default.html

http://nptel.ac.in/courses/104,106,063/Module%205/Lectures%208-10/Lectures%208-10.pdf

http://nptel.ac.in/courses/104101006/downloads/lecture-notes/mod10/lec1.pdf

http://nptel.ac.in/courses/104106064/lectures.pdf

http://web.uvic.ca/~djberg/Chem324/Chem324-16.pdf

http://classes.uleth.ca/201103/chem4000b/Substitution%20Reactions.pdf

http://chem.yonsei.ac.kr/chem/upload/CHE3103-01/122447755644547.pdf

http://www.mpilkington.com/Lecture_12.pdf

http://www.trentu.ca/chemistry/chem321h/lectures/lecture-321reactions.pdf

CHAPTER 6

Reactions in Square Planar Complexes

Contents

1. INTRODUCTION

The d^8 metal ions, such as Pt(II), Pd(II) and Ni(II), often form square planar complexes. The square planar complexes of Pt(II) are of particular interest in kinetic studies due to their high stability, ease of synthesis and moderate rates of reaction that enable the monitoring of the reaction. The area of discussion in these complexes is restricted only to the substitution reactions. As compared to the octahedral complexes, the crowding around the metal ion is less in square planar complexes. This is one of the important reasons that most of the substitution reactions in these complexes follow the SN^2 (associative mechanism).

2. THE TRANS EFFECT AND ITS APPLICATIONS

2.1 Trans Effect

Consider a substitution reaction shown in Figure 1.

Essentials of Coordination Chemistry
http://dx.doi.org/10.1016/B978-0-12-803895-6.00006-9
161

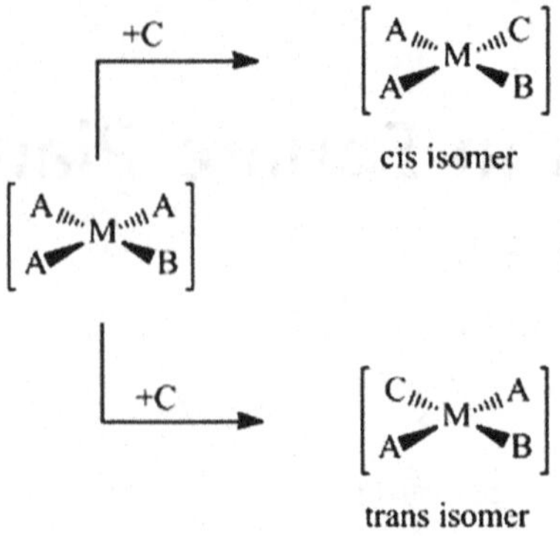

Figure 1 A ligand substitution reaction in square planar complex showing two possible isomeric products.

In this reaction, two isomeric products are possible. However, in practice, only one is formed. Certain ligands are capable of labilizing the ligands trans with respect to their position. This induces a facile removal of the labilized ligand and directs the incoming ligand at a trans position with respect to itself. This trans effect is the reason behind the formation of only one isomeric product. The ability of ligands to direct the incoming group at a position that is trans with respect to itself is called trans-directing ability.

An approximate order of ligands according to their trans-directing ability is

$$CN^-, CO, NO, C_2H_4 > PR_3, H^- > CH_3^-, C_6H_5^-, SC(NH_2)_2,$$
$$SR_2 > SO_3H^- > NO_2^-, I^-, SCN^- > Br^- > Cl^- > py > NH_3 > OH^-$$
$$> OH_2$$

The higher end of the series is occupied by ligands, such as CN^-, CO, NO, C_2H_4, which are characterized by high π-acidity (the ability of the ligands to accept the electron density back-donated by the metal ion into their low lying vacant π-orbitals).

From amongst the ligands without π-acidity, the most polarizable ones show high trans-directing ability, e.g. $I^- > Br^- > Cl^-$.

2.2 Applications of Trans Effect

The knowledge of trans effect while carrying out substitution reactions in square planar complexes has provided two interesting applications:
1. Selective synthesis of desired isomers out of various possible ones
2. Identification of geometrical isomers

2.2.1 Selective Synthesis

2.2.1.1 By Changing the Starting Material

From the trans-directing series, it is seen that Cl^- has higher trans effect than NH_3 ligand. Thus, for the synthesis of $[Pt(NH_3)_2Cl_2]$, if we select $[PtCl_4]^{-2}$ as the starting material and replace the two chloro ligands by ammine ligands sequentially, the first ammine ligand occupies any of the four positions. This is because all the positions are trans with respect to a chloro ligand. For deciding the position of the second incoming ligand, the trans-directing ability of Cl^- and NH_3 should be compared. The chloro that is more capable of directing the incoming ligand will force the incoming ammine ligand at a trans position with respect to itself, resulting into the formation of a cis-isomer of $[Pt(NH_3)_2Cl_2]$, as shown in Figure 2.

On the other hand, if we start with $[Pt(NH_3)_4]^{+2}$, a trans-isomer is obtained. Once again, the first incoming chloro ligand can occupy any of the four positions, but the second chloro ligand will occupy a position trans with respect to the chloro ligand, as shown in Figure 3.

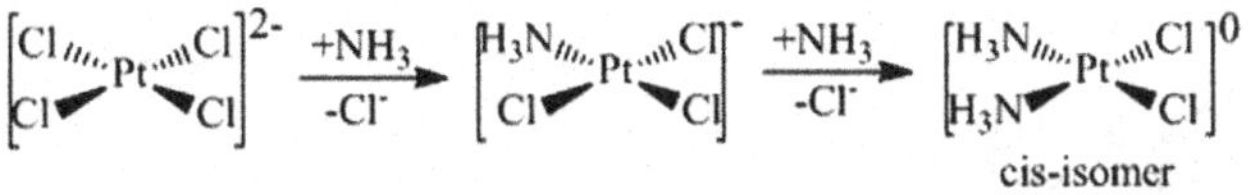

Figure 2 Starting with a complex bearing higher trans effect so as to produce a cis-isomer.

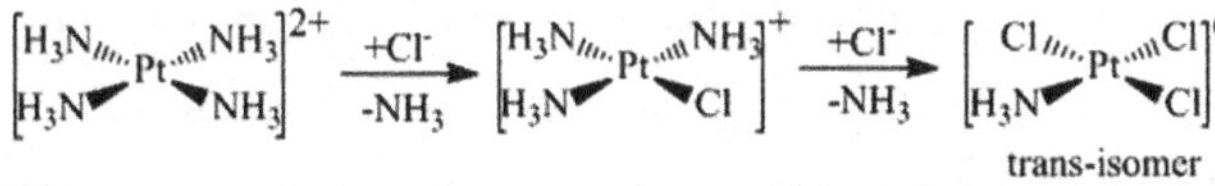

Figure 3 Starting with a complex bearing low trans effect so as to produce a trans-isomer.

Thus, in the preparation of complexes of type $[Ma_2b_2]$, we can generalize that starting with a complex with ligands having lower trans effect yields a trans-isomer. While starting with a complex with ligands having higher trans effect yields a cis-isomer.

2.2.1.2 By Changing the Sequence of Substitution

The order of trans-directing abilities of $NH_3 < Cl^- < NO_2^-$ permits the selective synthesis of isomers of the type $[Ma_2bc]$ from $[Ma_4]$ type of complexes. Consider the preparation of $[Pt^{+2}Cl_2(NO_2)(NH_3)]$ from $[PtCl_4]^{-2}$. If the ammonia ligand is introduced in the first step, it can occupy any of the four positions replacing a chloro ligand. In the second step, the incoming NO_2^- is offered a position trans with respect to chlorine due to its higher trans-directing ability as compared to ammonia. This results into the formation of a cis-isomer of the product. Instead of this, if it is decided to enter the NO_2^- in the first step, once again it can occupy any of the four positions, but the position of the ammine ligand to be entered in the second step will be determined on the basis of the trans-directing ability of Cl^- and NO_2^- ligands. The NO_2^- with higher trans effect of the two will labilize the Cl^- ligand opposite to it and will force the entry of the ammine group at a trans position with respect to itself. This will result into the formation of a trans-isomer.

Thus, in the preparation of complexes of type $[Ma_2bc]$, we can generalize that introduction of the ligand with lower trans effect in the first step results in the formation of a cis-isomer. While introducing the ligand with higher trans effect in the first step yields a trans-isomer.

The reactions concerning the above discussion are shown in Figure 4.

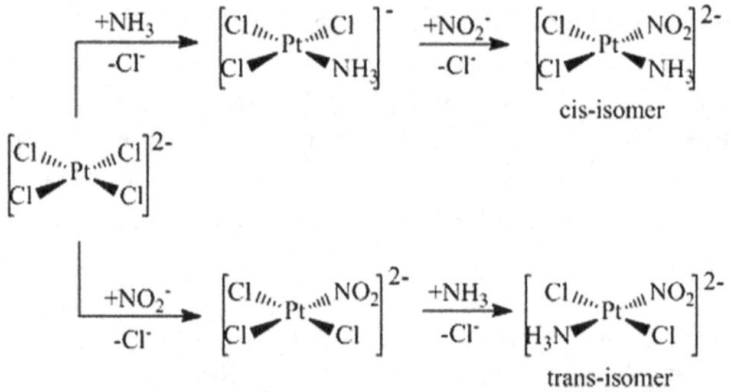

Figure 4 Changing the sequence of substitution to obtain a desired isomer.

2.2.1.3 By Changing the Starting Material, as Well as the Sequence of Substitution

From the above discussion, we have seen that it is possible to prepare a desired isomer by either changing the starting material or altering the sequence of substitution. Three geometrical isomers of the complexes of the type $[Mabcd]$ can be obtained by combining the above two methods. The synthesis of the three isomers of a complex $[Pt(py)(NH_3)ClBr]$ is shown in Figure 5. Here, the trans-directing ability of $NH_3 < py < Cl^- < Br^-$ and the Pt—N bond is stronger than the Pt—Cl bond.

2.2.2 Identification of Geometrical Isomers

Kurnakov used the phenomena of trans effect in distinguishing the cis and trans-isomers of square planar complexes of $[PtA_2X_2]$ type by treating them

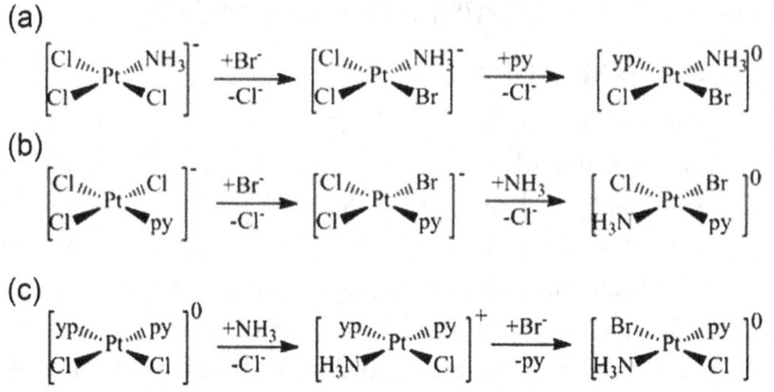

(a)

$$\begin{bmatrix} Cl_{\prime\prime\prime\prime\cdot}Pt\cdot^{\prime\prime\prime}NH_3 \\ Cl\diagup^{}\diagdown Cl \end{bmatrix}^{-} \xrightarrow[-Cl^-]{+Br^-} \begin{bmatrix} Cl_{\prime\prime\prime\prime\cdot}Pt\cdot^{\prime\prime\prime}NH_3 \\ Cl\diagup^{}\diagdown Br \end{bmatrix}^{-} \xrightarrow[-Cl^-]{+py} \begin{bmatrix} yp_{\prime\prime\prime\prime\cdot}Pt\cdot^{\prime\prime\prime}NH_3 \\ Cl\diagup^{}\diagdown Br \end{bmatrix}^{0}$$

(b)

$$\begin{bmatrix} Cl_{\prime\prime\prime\prime\cdot}Pt\cdot^{\prime\prime\prime}Cl \\ Cl\diagup^{}\diagdown py \end{bmatrix}^{-} \xrightarrow[-Cl^-]{+Br^-} \begin{bmatrix} Cl_{\prime\prime\prime\prime\cdot}Pt\cdot^{\prime\prime\prime}Br \\ Cl\diagup^{}\diagdown py \end{bmatrix}^{-} \xrightarrow[-Cl^-]{+NH_3} \begin{bmatrix} Cl_{\prime\prime\prime\prime\cdot}Pt\cdot^{\prime\prime\prime}Br \\ H_3N\diagup^{}\diagdown py \end{bmatrix}^{0}$$

(c)

$$\begin{bmatrix} yp_{\prime\prime\prime\prime\cdot}Pt\cdot^{\prime\prime\prime}py \\ Cl\diagup^{}\diagdown Cl \end{bmatrix}^{0} \xrightarrow[-Cl^-]{+NH_3} \begin{bmatrix} yp_{\prime\prime\prime\prime\cdot}Pt\cdot^{\prime\prime\prime}py \\ H_3N\diagup^{}\diagdown Cl \end{bmatrix}^{+} \xrightarrow[-py]{+Br^-} \begin{bmatrix} Br_{\prime\prime\prime\prime\cdot}Pt\cdot^{\prime\prime\prime}py \\ H_3N\diagup^{}\diagdown Cl \end{bmatrix}^{0}$$

Figure 5 Three isomeric products obtained by changing the starting material, as well as the sequence of substitution.

with thiourea (th). Here, the Pt—N bond is stronger than the Pt—Cl bond, and the trans-directing ability of thiourea is greater than NH_3.

While reacting a cis-isomer of $[Pt(NH_3)_2Cl_2]$ with thiourea, the two weakly bonded chloro ligands are replaced by thiourea in the first step. In the second step, the thiourea with higher trans-directing ability, as compared to ammonia, guides two more thiourea molecules to occupy trans positions with respect to themselves. Thus, all the four ligands on the substrate get substituted by thiourea, giving $[Pt(th)_4]$ as shown in Figure 6.

While in the case of the trans-isomer of $[Pt(NH_3)_2Cl_2]$, the two weakly bonded chloro ligands are replaced by thiourea in the first step, but in the second step, the higher trans effect of thiourea does not allow the ammine ligands to be sufficiently reactive so as to be replaced by the remaining thiourea ligands. Thus, the trans-isomer does not permit full substitution by thiourea and gives $[Pt(NH_3)_2(th)_2]$, as shown in Figure 7.

Other than thiourea, the thiosulfate $(S_2O_3)^{-2}$ ligand also shows similar reactions with the complexes of type $[PtA_2X_2]$.

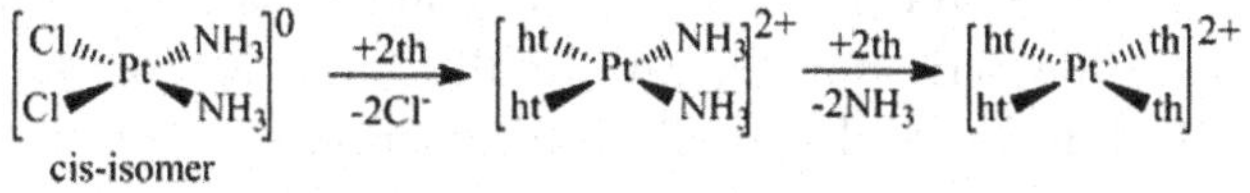

$$\begin{bmatrix} Cl_{\prime\prime\prime\prime\cdot}Pt\cdot^{\prime\prime\prime}NH_3 \\ Cl\diagup^{}\diagdown NH_3 \end{bmatrix}^{0} \xrightarrow[-2Cl^-]{+2th} \begin{bmatrix} ht_{\prime\prime\prime\prime\cdot}Pt\cdot^{\prime\prime\prime}NH_3 \\ ht\diagup^{}\diagdown NH_3 \end{bmatrix}^{2+} \xrightarrow[-2NH_3]{+2th} \begin{bmatrix} ht_{\prime\prime\prime\prime\cdot}Pt\cdot^{\prime\prime\prime}th \\ ht\diagup^{}\diagdown th \end{bmatrix}^{2+}$$
cis-isomer

Figure 6 Identification of cis-isomer using Kurnakov's reaction.

$$\begin{bmatrix} H_3N_{\prime\prime\prime\prime\cdot}Pt\cdot^{\prime\prime\prime}Cl \\ Cl\diagup^{}\diagdown NH_3 \end{bmatrix}^{0} \xrightarrow[-2Cl^-]{+2th} \begin{bmatrix} H_3N_{\prime\prime\prime\prime\cdot}Pt\cdot^{\prime\prime\prime}th \\ th\diagup^{}\diagdown NH_3 \end{bmatrix}^{+} \xrightarrow[\text{Not possible}]{+th} \begin{bmatrix} H_3N_{\prime\prime\prime\prime\cdot}Pt\cdot^{\prime\prime\prime}th \\ ht\diagup^{}\diagdown NH_3 \end{bmatrix}^{2+}$$
trans-isomer

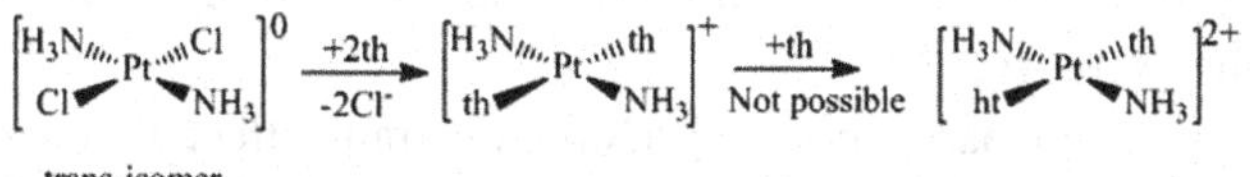

Figure 7 Identification of trans-isomer using Kurnakov's reaction.

3. THEORIES FOR EXPLAINING TRANS EFFECT

3.1 Chatt and Orgel Theory

This theory successfully explains the high trans-directing ability of ligands with π-acidity. The ligands such as CN^-, CO, NO, C_2H_4, PR_3 have an ability to accept electron density back-donated by the metal ion into their low-lying vacant π-orbitals. In the square planar complexes of Pt(II), such ligands accept the electron density from the filled platinum orbitals (dyz) into their suitable vacant pz or d-orbitals. This results into the formation of a π-bond, which increases the electron density between the metal ion and π-acidic ligand (L). The electron density at the bond opposite to this ligand is correspondingly decreased, which lengthens this bond and in turn creates a space for the entry of the incoming ligand. Finally, the incoming ligand (Y) occupies a position trans with respect to the π-acidic ligand (L) via a five-coordinated intermediate, as shown in the figure.

Diagrams showing the overlapping of the metal-ligand orbitals to form π-bonds are shown in Figure 8.

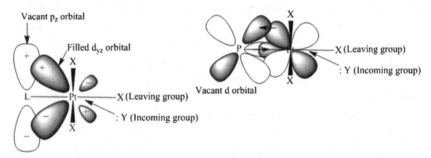

Figure 8 A dπ–pπ bond in the activated complex formed during the substitution reaction in complex PtLX$_3$ (left). Double-bond formation in R$_3$P and Pt(II) (right).

The transfer of electron density from the metal orbitals by the π-acidic ligand L facilitates the entry of the incoming group (Y) and expedites the reaction. Otherwise, the formation of π-bond between the metal and ligand reduces the activation energy for the formation of the activated complex, which increases the speed of the reaction.

3.2 Grinberg's Polarization Theory

According to this theory, the trans-directing ability of ligands with greater polarizability is more.

For a square planar complex [PtX$_4$], the trans effect is not operative. Here, the positive charge on the central metal ion induces dipole in the

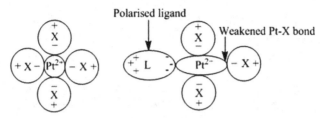

Figure 9 Cancellation of charges in Pt(II) square planar complex, PtX$_4$ (left). Weakening of Pt—X trans to L in a Pt(II) square planar complex, PtX$_3$L (right).

ligands, as shown in the Figure 9. Since these dipoles are equal and opposite, they are mutually cancelled and no net trans effect is observed.

However, in the cases of complexes of type [PtX$_3$L], as shown in the Figure 9, the induced dipoles of the two 'X' ligands occupying the trans position get cancelled, being equal and opposite. But the induced dipoles for a highly polarizsable ligand L occupying a position trans with respect to a ligand 'X' remains non-zero. The ligand L now induces a dipole in the Pt(II) in such a way that the positive charge on Pt(II) in the regions opposite to L reduces. This in turn reduces the attraction of Pt(II) for the 'X' ligand opposite to L. This facilitates the replacement of the 'X' ligand trans to L.

The electrostatic polarization theory predicts that the trans effect is observed only if the central metal ion of the complex is polarizsable.

4. MECHANISM OF SUBSTITUTION REACTIONS

As compared to the octahedral complexes, the crowding around the metal ion is less in square planar complexes. Due to this, an associative mechanism (SN^2) is known to be followed in the substitution reactions involving square planar complexes. An associative mechanism with or without solvent intervention is possible. A mechanism involving the initial coordination of a solvent molecule to replace the leaving group followed by the replacement of a solvent molecule by an entering ligand is called a solvent path. In a path known as reagent path, there is no solvent intervention.

The ligand substitution reactions of square planar complexes of Pt(II) of the type PtA$_2$XL are stereospecific. If the substrate is a cis-isomer, the product is invariably a cis-isomer and vice versa. The reaction occurs via trigonal bipyramidal intermediate, as shown in Figure 10.

A mechanism involving solvent (H$_2$O) intervention, shown in Figure 11, is possible in a substitution reaction on a square planar complex of type [PtA$_3$X] when performed in an aqueous medium. For the reaction

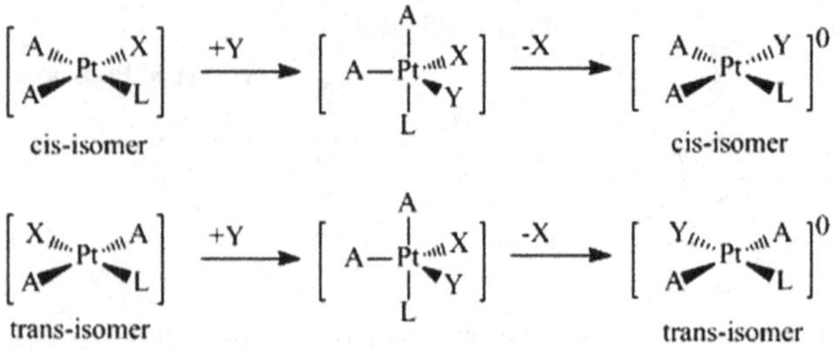

Figure 10 SN^2 mechanism for substitution reactions in cis- and trans-isomers of a square planar complex.

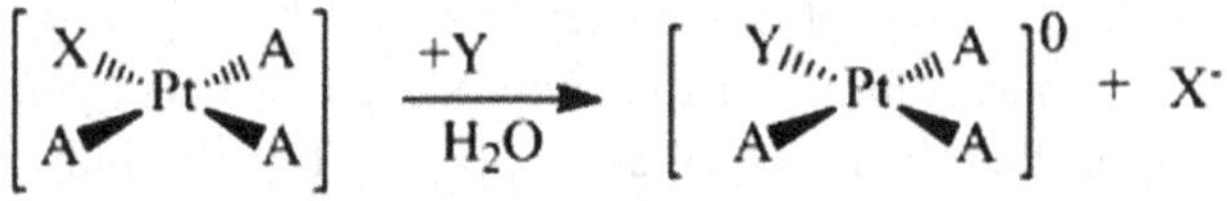

Figure 11 A mechanism involving solvent intervention in a substitution reaction on a square planar complex of type [PtA$_3$X].

The rate $= K_1 [PtA_3X] + K_2 [PtA_3X] [Y]$

Where K_1 and K_2 are first and second order rate constants corresponding to solvent path and reagent path.

The analysis of rate constants can be done using a large concentration of nucleophile (Y). Here, the observed rate constant K_{obs} is a pseudo first-order rate constant, which can be given as

$$K_{obs} = K_1 + K_2[Y]$$

(This equation is like $y = mx + c$. From x versus y plot following this type of equation, m is the slope of the line and c is the intercept on y-axis.)

A plot of nucleophile concentration [Y] against pseudo first-order rate constant (K_{obs}) for substitution reactions of various ligands with trans-[Pt(py)$_2$Cl$_2$] is shown in Figure 12.

From this plot, it is seen that the intercept (K_1) for different ligands remains the same. However, the slope (K_2) is found to vary with different ligands.

The equation for rate suggests that a substitution reaction on [PtA$_3$X] to give [PtA$_3$Y] follows a two-path mechanism, as shown in Figure 13.

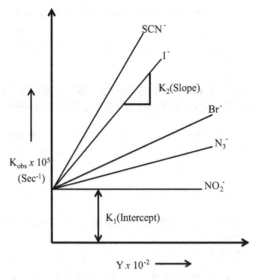

Figure 12 A plot of nucleophile concentration against pseudo first-order rate constant for substitution reactions of various ligands with trans- $[Pt(py)_2Cl_2]$.

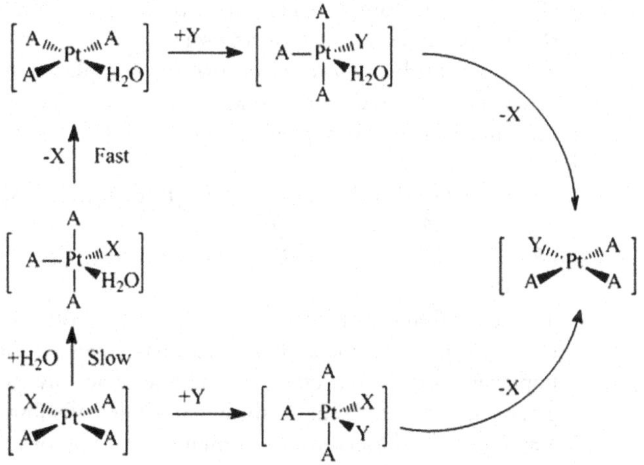

Figure 13 A two-path mechanism for substitution reaction on $[PtA_3X]$ to give $[PtA_3Y]$.

The path shown in the clockwise direction is a solvent-intervention path. The solvent path follows SN^2 mechanism, as shown in the figure. The path shown in the anti-clockwise direction is a reagent path.

5. FACTORS AFFECTING THE RATE OF SUBSTITUTION REACTIONS IN SQUARE PLANAR COMPLEXES

The observed rate of substitution reactions in square planar complexes is the result of multifaceted interplay between several factors. The effect of some such factors is summarized in Table 1.

Table 1 Comments on certain factors affecting the rate of substitution reactions in square planar complexes

Factor	Comments
Charge on the complex	An increase in positive charge on the complex increases the difficulty in removing the outgoing ligand, which could decrease the rate of reaction. However, it seems that this decrease in rate of reaction is nearly counterbalanced by increased attraction towards the incoming nucleophile. The complexes $[PtCl_4]^{-2}$, $[Pt(NH_3)Cl_3]^{-1}$, $[Pt(NH_3)_2Cl_2]$ and $[Pt(NH_3)_3Cl]^{+1}$ have progressively increasing positive charge on the complex. However, the rate of hydrolysis for all these complexes is almost the same.
Trans effect	The rate of substitution reaction of the complexes of type $[Pt(NH_3)Cl_2L]$ with pyridine as the incoming ligand shows the following result trans effect $C_2H_4 > NO_2^- > Br^- > Cl^-$ relative rates $100 > 90 > 3 > 1$
Leaving group	For a reaction $[Pt(dien)X] + py \rightarrow [Pt(dien)py] + X$ The rate varies with the ligand X as $NO_3^- > H_2O > Cl^- > Br^- > I^- > N_3^- > SCN^- > NO_2^- > CN^-$
Solvent	The identification of solvent path for the substitution reactions in square planar complexes has increased the importance of solvent effect on the rate. The solvents with greater coordinating ability show faster rates of substitution reactions. The observed rate constants for some of the popular solvents vary as $ROH < H_2O < CH_3NO_2 < (CH_3)_2SO$

6. EXERCISES

6.1 Multiple Choice Questions

1. Which one of the following has the highest trans-directing ability?

    (a) NO_2^- (b) Br^-
    (c) CN^- (d) Cl^-

2. Which one of the following is the correct order of decreasing trans effect?

    (a) $OH^- > Br^- > C_6H_5^- > H^- > NO$ (b) $Br^- > C_6H_5^- > NO > OH^- > H^-$
    (c) $C_6H_5^- > NO > Br^- > OH^- > H^-$ (d) $NO > H^- > C_6H_5^- > Br^- > OH^-$

3. Substitution reactions of cis and trans PtA_2LX with Y to yield PtA_2LY is

    _____.

    (a) stereospecific   (b) stereoselective
    (c) not possible   (d) unknown

4. The electrostatic polarization theory can explain the trans effect of the ligands lying at _____ of the trans effect series.

    (a) higher end   (b) lower end
    (c) middle   (d) alternate positions

6.2 Short/Long Answer Questions

1. Give the steps involved in the formation of cis and trans-isomers by treating $[Pt^{II}Cl_4]^{2-}$ ion with NH3.
2. How can you distinguish between cis and trans-isomers of $[PtA_2X_2]^0$ type complexes?
3. What is trans effect? Explain the applications of "trans effect" using synthesis of Pt(II) complexes.
4. Discuss the electrostatic polarization theory for trans effect.
5. Discuss the mechanism of substitution reaction in Pt(II) square planar complexes.

SUGGESTED FURTHER READINGS

Many books on advanced coordination chemistry offer the contents of this chapter. In addition to such books, a reader may refer to the following websites to get further information and knowledge.

http://web.uvic.ca/~mcindoe/423/Chem324-17.pdf
http://www.chem.ox.ac.uk/icl/dermot/mechanism1/lecture1/ligands.html

http://www.adichemistry.com/inorganic/cochem/reactionmechanism/transeffect/trans-effect-1.html
http://chemwiki.ucdavis.edu/?title=Inorganic_Chemistry/Organometallic_Chemistry/Fundamentals/The_trans//cis_Effects_%26_Influences
http://www.mpilkington.com/Lecture_14.pdf

CHAPTER 7

Basic Organometallic Chemistry

Contents

1. INTRODUCTION

Organometallic chemistry involves the creation of compounds with C—M bonds with $C^{-\delta}$—$M^{+\delta}$, as compared to that of an organic compound, which contains nonpolar C—C bonds. The partial negative charge on such carbon of an organometallic compound is the origin of its majority of applications, as it invokes the tendency to make a nucleophilic attack in that carbon and, in turn, facilitates a wide variety of organic reactions.

Essentials of Coordination Chemistry
http://dx.doi.org/10.1016/B978-0-12-803895-6.00007-0
173

Compounds with at least one bond between metal and carbon of an organic group are known as organometallic compounds. Hence, compounds like $Ti(OC_4H_9)_4$, $Ca[N(CH_3)_2]_2$ and $Fe(SC_5H_{11})_3$ are not organometallic compounds, but $C_6H_5Ti(OC_4H_9)_3$ is an organometallic compound. Almost all the elements in the periodic table, other than noble gases, form compounds with organic-carbon bonded to them. This includes nonmetallic elements such as boron, silicon, arsenic and selenium, etc. Thus, the 'metallic' in organometallic does not actually stand for 'metal'; it refers to all the elements having less electronegativity than carbon (2.5). Based on this, the derivatives of nitrogen (3.0), oxygen (3.5), sulphur (2.6) and fluorine (4.0), etc. are not regarded as organometallic compounds. Traditionally, the compounds like CaC_2, $Hg(CNO)_2$ and $Fe(CN)_6^{-4}$ are also not regarded as organometallic compounds. 'Cacodyl', $As_2(CH_3)_4$ (tetramethyldiarsine), a reddish-brown fuming liquid with a bad smell, was the first organometallic derivative isolated in 1760.

2. NATURE AND TYPES OF METAL–CARBON BONDING

Essentially, the five types of bonds listed below are observed between metal and carbon in organometallic compounds. Many organometallic compounds have more than one type of bond in them.

1. Metal–carbon ionic bonds
2. Metal–carbon two-centre two-electron covalent bonds
3. Metal–carbon three-centre two-electron bridge bonds
4. Metal–carbon π-bonds
5. Metal–carbon multiple bonds

Primarily, the electronegativity of the metal involved in the organometallic compounds decides the percentage of ionic character in the metal–carbon bonds. The elements with their positions farther from carbon in the periodic table form essentially ionic bonds. Whereas the elements relatively near to carbon in the periodic table give metal–carbon bonds with less ionic character. In the case of organometallic oligomers, if an electron-deficient metal is involved, metal–carbon three-centre two-electron bridge bonds similar to that of diborane are observed. Metal–carbon π-bonds are observed in the organometallic compounds that contain the organic group with high π-acidity (the ability to donate π-electrons). Organic compounds with one or more π-bonds can also donate the π-electron density to form manifold bonds, as in Zeise's salt and ferrocene kind of molecules.

3. PREPARATION OF METAL–CARBON BONDS

3.1 Oxidative-Addition Reactions

Metals with high electropositivity readily get oxidized by the addition of a group like 'RX'. The following equations represent one- and two-electron oxidative addition reactions respectively:

$$2M + R-X \rightarrow M-R + M-X-R$$

$$M + R-X \rightarrow R-M-X$$

Representative examples of these two types of reactions are

$$2Na + R-Cl \rightarrow Na-R + Na-Cl \quad (R = alkyl \; or \; aryl)$$

With highly electropositive alkali metals, metal alkyl formed in the reaction undergoes the Wurtz–Fittig reaction as $Na-R + R-X \rightarrow R-R + Na-Cl$ to give an alkane and a salt.

This problem is avoided by the use of finely grounded metals and efficient stirring that enhances the formation of metal alkyls. A very slow and controlled addition of alkyl halides also reduces the extent of the side reaction.

The alkyls of relatively less electropositive metals are generally prepared using the metal in the form of an alloy:

$$2C_2H_5I + 2Hg/Na \rightarrow Hg(C_2H_5)_2 + 2NaI$$

Grignard reagents and related compounds are prepared in ether by two-electron oxidative–addition reactions as $R-Br + Mg \rightarrow Mg-R-Br$.

A majority of the catalytic cycles involving transition metal organometallics include a step of oxidative addition, leading to the formation of a six-coordinate saturated complex as

$$IrCl(CO)(PPh_3)_2 + CH_3I \rightarrow IrCl(I)(CH_3)(CO)(PPh_3)_2$$

3.2 Transmetallation

These reactions, also known as metal–metal exchange reactions, are represented as

$$M + M'-R \rightarrow M-R + M'$$

The above reaction is a redox reaction in which M gets oxidized while the M' gets reduced. The tendency to get oxidized is the same as the tendency to lose electrons. Thus, metals with greater electropositivity readily play the role of M in the transmetallation reaction. Thus, the following reactions logically represent transmetallation:

$$2Li + R{-}Mg{-}R \rightarrow 2Li{-}R + Mg$$

$$2Na + R{-}Hg{-}R \rightarrow 2Na{-}R + Hg$$

$$M + HgR_2 \rightarrow MR_2 + Hg \quad (M = Ca, Sr, Ba)$$

3.3 Carbanion–Halide Exchange Reactions

This type of reaction involves the interchange of the carbanion (alkyl/aryl) and halide groups present on two different metals and can be represented as

$$MX_n + M'R_m \rightarrow MX_{n-a}R_a + M'R_{m-a}X_a$$

These reactions proceed in the direction leading to the bonding of the halogen with more electropositive metal and of the carbanion with less electropositive metal as

$$BeCl_2 + 2LiR \rightarrow BeR_2 + 2LiCl$$

$$MgBr_2 + LiR \rightarrow Mg{-}R{-}Br + LiBr$$

$$AlCl_3 + 3LiR \rightarrow AlR_3 + 3LiCl$$

$$ZrCl_4 + 4Li(CH_2){-}Si{-}(CH_3)_3 \rightarrow Zr(CH_2{-}Si{-}(CH_3)_3)_4 + 4LiCl$$

3.4 Metallation

These reactions, also known as metal–hydrogen exchange reactions, may be represented as the following:

$R{-}H + M{-}R' \rightarrow M{-}R + R'{-}H$ have a reversible nature, with the equilibrium shifted largely on the product side. This equilibrium can be further pushed in many cases to achieve the complete metallation.

A nucleophilic attack of the metal bonded carbon of the organometallic compound on a hydrogen atom of the organic substrate affords the reaction. The hydrogen atom undergoing metallation must therefore be quite acidic, while the organometallic compound should have a strong carbanionic character. Some examples of metallation reactions are represented as follows:

$$(C_6H_5)_3CH + LiC_2H_5 \rightarrow LiC(C_6H_5)_3 + C_2H_6$$

$$C_6H_5CH_3 + KC_6H_5 \rightarrow KCH_2C_6H_5 + C_6H_6$$

3.5 Insertion

This method for the preparation of organometallic compounds involves insertion of an alkene into a metal–hydrogen bond. A general insertion reaction may be represented as follows:

$$R_3-Sn-N-R_2 + CO_2 \rightarrow R_3-Sn-OC(O)-NR_2$$

The insertion of an alkene in a metal–hydrogen bond results in the formation of a bond between metal and carbon as

Some important insertion reactions in organometallic chemistry are

$$B_2H_6 + 6RCH{=}CH_2 \rightarrow 2B(CH_2CH_2R)_3 \text{ (Hydroboration)}$$

$$LiAlH_4 + 4CH_2{=}CH_2 \xrightarrow[\text{pressure}]{100^\circ C} Li[Al(CH_2CH_3)]_4]$$

$$(C_6H_5)_3-Si-H + RCH{=}CH_2 \rightarrow Si-(CH_2CH_2R)-(C_6H_5)_3$$

$$SiHCl_3 + RC{\equiv}CH \rightarrow Cl_3SiCH{=}CHR \rightarrow Cl_3SiCHRCH_2SiCl_3$$

3.6 Methylenation

Introduction of a methylene group using diazomethane in ether produces metal–carbon bonds as follows:

$$MCl_4 + CH_2N_2 \rightarrow Cl_3-M-(CH_2Cl) + N_2$$

This reaction can be extended to a replacement of all the M–Cl bonds by M–C bonds, but the reaction becomes progressively difficult. A representative illustration for such a reaction is provided as

$$Cl_2-Sn-(CH_3)_2 + CH_2N_2 \rightarrow Cl-Sn-(CH_2Cl)\ (CH_3)_2 + N_2$$

4. CLASSIFICATION OF ORGANOMETALLIC COMPOUNDS

The organometallic compounds can be functionally classified on the basis of the nature of M–C bond [1]. The extent of polarity in the bond can explain the structure and reactivity of the organometallic compounds.

The main group elements with much greater electropositivity than carbon tend to form ionic organometallic compounds, whereas those with the comparable electropositivity as carbon form essentially covalent organometallic compounds. The transition and the inner-transition metals exhibit a variety of bonding in organometallic compounds.

4.1 Ionic Organometallic Compounds

1. Colourless except those derived from aromatic nucleus or conjugated systems
2. Highly reactive with air, water, protic reagents and halogenated compounds
3. Nonvolatile solids
4. Insoluble in hydrocarbon solvents
5. Decompose on heating without melting
6. The stability of the corresponding carbanions generally governs the stability of these compounds

Compounds containing unstable anions like C_nH_{2n+1} are very reactive and difficult to isolate. While the metal derivatives like $Na^+{}^-C(C_6H_5)_3$ and $Ca^{2+}(C_5H_5)_2{}^{2-}$ are more stable, attributed to the higher stability of the carbanions.

Table 1 summarizes the pKa value of H-R for some organic groups. The organic groups with a smaller value of pKa of H-R provide greater carbanion stability and, in turn, greater stability to the ionic organometallic compound.

4.2 σ-Bonded Organometallic Compounds

Elements with electronegativity greater than one form organometallic compounds containing two-centre two-electron bonds with carbon. The bonds obviously contain considerable ionic character. These compounds have generally inherited the properties of organic compounds. The difference in the properties of these compounds is observed due to the following factors:

1. Bond polarity
2. Electronuclear effects of the metal
3. Electronic and steric effects of the organic group
4. Solvent effects

4.2.1 Bond Polarity

Because of the higher electronegativity of carbon, the electron density in M−C bonds in such compounds is more towards carbon ($M^{+\delta}-C^{-\delta}$).

Table 1 pK_a of R–H in some organic groups

Name of organic group	pKa of R–H
Cyclopentyl (C_5H_9)	44.1
Isopropyl (C_3H_7)	43.2
Methyl (CH_3)	38.9
Vinyl ($CH_2{=}CH$)	37
Phenyl (C_6H_5)	36.9
Pentachlorophenyl (C_6Cl_5)	30.5
Pentafluorophenyl (C_6F_5)	23
Cyclopentadienyl (C_5H_6)	18

From the Pauling's equation for calculating the percentage of ionic character in a bond, it has been observed that the Si–C, Al–C, Mg–C and Na–C bonds have approximately 12, 22, 34 and 45% ionic characters, respectively. Merely 6% ionic character in C–Cl bonds, in the case of chloro- organic compounds, affords many reactions, which indicates that even a small ionic character in this bond affects the reactivity of compounds. However, the increase in the stability of carbanion 'R–' does not increase the reactivity of organometallic compounds of a particular metal.

4.2.2 Electronuclear Effects of the Metal

Lewis acidity in the metal alkyls arises due to the presence of low-energy vacant orbitals in the metals. These low-lying orbitals, in the case of the main group metallic elements (I, II and III), are the np orbitals. The metalloid elements of the main group (IV, V and VI) and some transition metals have n and $(n - 1)$d vacant orbitals. The Lewis acidity increases in the main group while moving from group I to III. Thus, the order of Lewis acidity in these alkyls is $NaR < MgR_2 < AlR_3$. The organometallic compounds, with the formula MR_n of main group IV to VI, have much less Lewis acidity and hence are not readily separated as adducts with Lewis bases. The derivative with the general formula $MR_{n-a}X_a$ may have higher Lewis acidity, provided that the group X is more electronegative than R. This is due to the contraction of metal d-orbitals due to an increased positive charge upon the introduction of more electronegative species 'X'. This, in turn, allows a facile participation of the d-orbitals with the donor atoms.

The organometallic compounds of transition metals are found to be less stable than those of the main group elements. Since a large number of binary transition metal alkyls and aryls arise from coordinatively unsaturated transition metals, their thermodynamic decomposition is easy to occur. The coordinatively unsaturated transition metals afford stable complexes by coordinating with Lewis bases such as R_2O, R_2S, R_3P and X^-. Otherwise, the filled $(n-1)$d-orbitals of the transition metals with suitable symmetry can overlap with the available anti-bonding π-orbitals of groups like carbonyl, vinyl or phenyl. This type of back-bonding releases the increased electron density from the metal centres and enhances the electron density between the metal and carbon centres.

4.2.3 Electronic and Steric Effects of Organic Moiety
The degree of association of aluminium alkyls AlR_3 in inert solvents follows the sequence; R=Me > Et > Pr > tert-Bu. This observation is attributed to the following:
1. Higher inductive effect in the branched alkyl groups, which increases the electron density on Al atoms
2. Steric effect of larger alkyl groups
   The unsaturation and its location in the organic group also play a significant role in the stability of M—C bonds in compounds like MR_n. For α, β unsaturation, the stability of the potential carbanion increases with the increase in 's' character of the α carbon atom as

$$M-CH-R\ (sp^3) < M-CH=CH-R;\ M-C_6H_5\ (sp^2) < M-C\equiv C-R\ (sp)$$

4.2.4 Solvent Effects
Donor solvents generally reduce molecular association between the metal and the carbon atom of the alkyl groups. The increased electron density around the metal atom, due to coordination with the solvent molecules with Lewis base nature like ethers and amines, enhances the polarization of the metal–carbon bond. Due to this, the carbanionic character of the organic groups gets enhanced.

5. NOMENCLATURE

A working solution for naming the organometallic compounds is provided in the following section. Organometallic compounds with mono-alkyl groups attached to the metals may be named as methyllithium for CH_3Li, while di/tri-alkyl compounds may be named as diethylzinc for $(C_2H_5)Zn$.

According to one of the systems, the organic groups/hydrogen atoms bonded to metal are named in alphabetical order, with no space between groups, followed by the name of the metal. The hydrogen attached to a metal is designated with the prefix, 'hydrido'. The numbers of identical complex groups are indicated by prefixes such as bis, tris, tetrakis and so on, as in $(Bu^i_2AlH)_3$, hydridobisisobutylaluminium.

The number of C atoms of an organic group bonded to metal are indicated by the prefix η, which is read as hapto, as in $(C_5H_5)_2Fe$ (named bis(η^5-cyclopentadienyl)iron). Further, the groups bridging two metal centres are given a prefix μ, as in $(Me_3Al)_2$; di-μ-methyl(tetramethyl) dialuminium.

6. FLUXIONAL ORGANOMETALLIC COMPOUNDS

The molecules that undergo intramolecular rearrangements that are detected in nuclear magnetic resonance (NMR) spectroscopy are termed as structurally nonrigid molecules [2]. If these intramolecular rearrangements involve all the interconverting species that are observable and are chemically, as well as structurally, equivalent, the molecules are defined as fluxional compounds. A significant number of fluxional organometallic compounds have been identified to date.

Consider an organometallic trigonal bipyramidal complex $Fe(CO)_5$. Three of the metal-bonded carbon monoxide molecules occupy the equatorial positions labelled 2, 3 and 4, whereas the remaining two occupy the axial positions labelled 1 and 5 in Figure 1. An intramolecular rearrangement without any net chemical or structural change is demonstrated by the exchange of two axial carbon monoxide ligands with two of the equatorial ones, as shown in Figure 1.

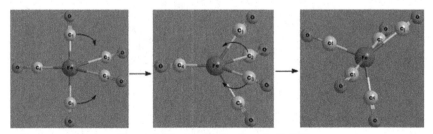

Figure 1 Fluxional behaviour in $Fe(CO)_5$.

This type of fluxional behaviour is exhibited by several organometallic compounds and is readily detected by the analysis of line shape in NMR [3]. The major limitation of this method is that it is useful only in the free energy change in the range of 40–80 kJ/mol. Microwave and infrared spectroscopy are also sometimes useful in studying the fluxional behaviour of organometallic compounds.

The line shape analysis of the NMR spectra of fluxional molecules conducted at low temperatures exhibit broadening of the lines due to the reduced rate of exchanges. The analysis also provides some information regarding the rate and mechanism of the exchange.

7. APPLICATIONS OF ORGANOMETALLIC COMPOUNDS

Organometallic compounds are applied widely as stoichiometric and catalytic reagents in preparing various organic compounds. The stoichiometric reagents are the reagents that are used as initiators and are not regenerated along with the product formation but are converted into some other species. The organometallic compounds of the main group are generally employed as stoichiometric reagents, illustrated by the role of alkyllithium in the polymerization of alkenes. The alkyllithium, as shown in Figure 2, gets converted into an inactive species, lithiumhydride, at the end of the process.

Moreover, these reagents are required in stoichiometric proportions, as they do not remain active at the end of the process in which they are used.

The transition metal organometallic compounds/complexes are extensively used as catalytic reagents in many organic reactions. These catalysts are sometimes in the same phase as the reactants. Such catalysis is known as homogeneous catalysis. When the catalysts and the reactants are in different phases, the catalysis is called heterogeneous catalysis. The role of transition

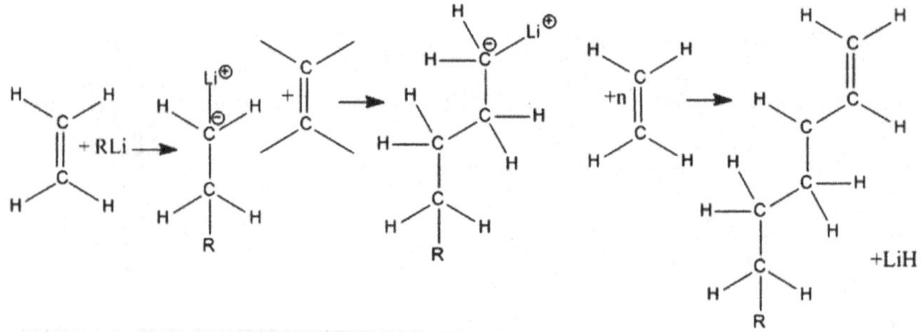

Figure 2 Alkyl lithium-initiated polymerization of ethylene.

metal carbonyl complexes in catalysis is discussed, with suitable examples in the chapter related to metal carbonyls.

8. SOME ORGANOMETALLIC COMPOUNDS OF INTEREST

8.1 Methyllithium (CH$_3$Li)

It can be readily prepared by a reaction between methyl bromide/methyl chloride and a suspension of lithium in diethyl ether as $2Li + CH_3Br \rightarrow LiCH_3 + LiBr$.

Methyllithium is a strong basic and a powerful nucleophile. Compared to another synthetically important organolithium compound, n-butyllithium, methyllithium reacts slowly with tetrahydrofuran at room temperature. An etheral solution of methyllithium is stable for a long time. Most of the reactions involving methyllithium are conducted at low temperatures. Methyllithium is used for deprotonations and as a source of methyl anion. Ketones can be converted to tertiary alcohols using methyllithium as

$$2C_6H_5COC_6H_5 + CH_3Li \rightarrow C_6H_5C(CH_3)OC_6H_5Li + H^+$$
$$\rightarrow C_6H_5C(CH_3)OHC_6H_5 + Li^+.$$

It is also used to convert halides into methyl compounds as $PCl_3 + 3CH_3Li \rightarrow P(CH_3)_3 + 3LiCl$.

It reacts with carbon dioxide to produce lithiumacetate as

$$CH_3Li + CO_2 \rightarrow CH_3COOLi.$$

Halides of transition metals react with methyllithium to give methyl compounds. This reaction is alternatively afforded by organocopper compounds like lithiumdialkylcuprates which are also known as Gilman reagents. These reagents are widely used for nucleophilic substitutions of epoxides, alkyl halides and for conjugate additions to α,β-unsaturated carbonyl compounds by methyl anion.

CH$_3$Li is the empirical formula for methyllithium, it exists in oligomeric forms both in solid, as well as solution, states. Tetrameric and hexameric forms of methyllithium have been identified using NMR and crystallography. The tetrameric form in Figure 3 shows that the four lithiums occupy alternate diagonal corners of the opposite faces of the cube, forming a tetrahedron. The four vacant corners of the cubes are also the alternate diagonal positions of the opposite faces of the cube, which are occupied by the carbons of the methyl group. The bond length measurements in the molecule indicate that the structure is a distorted cube.

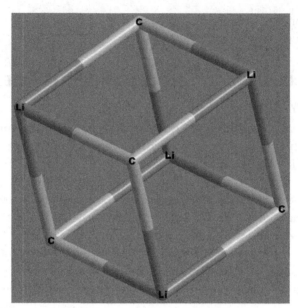

Figure 3 Tetrameric form of methyllithium.

This electron-deficient unit does not obey the octet rule, as it does not have enough electrons to form four two-centre two-electron bonds around each carbon atom. The interaction between an sp^3 hybrid orbital of a methyl group with three 2s orbitals of the Li atoms form a symmetric four-centre two-electron bonding orbital, which accounts for the bonding in this compound.

8.2 Zeise's Salt (K[PtCl$_3$(C$_2$H$_4$)]H$_2$O)

Potassium trichloro(ethene)platinate(II), also known as Zeise's salt, has the formula $K[PtCl_3(C_2H_4)]H_2O$ and contains η^2-ethylene ligand. The compound is a yellow-coloured coordination complex that is stable in air. The platinum atom has a square planar geometry. While reacting K_2PtCl_4 with ethanol in 1827, a Danish chemist Zeise synthesized this metal olefin complex containing a platinum-bound ethylene moiety, which incidentally represented the first metal—olefin complex.

The traditional approach of the preparation of Zeise's salts involves a reaction between $K_2[PtCl_4]$ and ethylene in the presence of a catalytic amount of stannous chloride as

$$K_2[PtCl_4] + C_2H_2 \xrightarrow{\text{SnCl}_2} K_2[PtCl_3(C_2H_4)]$$

A microwave synthesis of Zeise's salt requires 15 min of heating at 130 °C using K_2PtCl_4 as a starting material, a 1:1:1 ratio of water:-ethanol:concentrated hydrochloric acid as a solvent and a loading of 50 psi of ethylene [4].

Figure 4 shows bonding in metal-alkene complexes. The π-electrons from a double bond of the ethylene donate electrons to empty metal d-orbitals to form a σ-bond, as shown in Figure 4, and filled d-orbitals of the platinum form π-bonds with empty π^*-orbitals of a double bond. This mode of σ- and π-bond formation is synergic. A σ-bond is strengthened by π-bond formation due to the flow of electron density from platinum to ethylene, and π-bond formation is strengthened by σ-bond formation because of the flow of electron density from ethylene to platinum.

A square planar $[PtCl_3(C_2H_4)]$ molecule is shown in Figure 5.

Zeise's salt reacts with water to replace a chloro ligand by aquo, upon which base hydrolysis decomposes to give acetaldehyde as

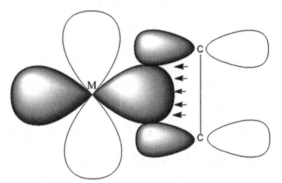

Empty d-orbital of metal   Filled π orbitals of ethylene

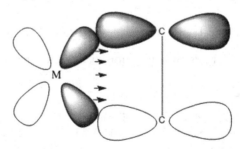

Filled d-orbital of metal   Empty π^* orbitals of ethylene

Figure 4 Synergic σ- and π-bond formation in Zeise's salt.

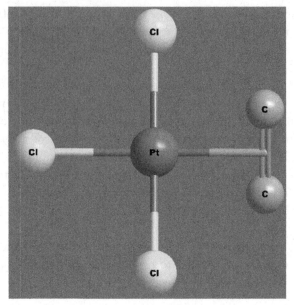

Figure 5 Structure of Zeise's salt.

$$K_2[PtCl_3(C_2H_4)] + H_2O \rightarrow K_2[PtCl_2(H_2O)(C_2H_4)]$$
$$\xrightarrow{OH^-} Pt + CH_3CHO + KCl + 2HCl$$

An attack by nucleophiles, such as cyano and carbonyl, results in the replacement of the alkene as in

$$K_2[PtCl_3(C_2H_4)] + CN^- \rightarrow K_2[Pt(CN)_4]$$
$$K_2[PtCl_3(C_2H_4)] + CO \rightarrow K_2[PtCl_3(CO)]$$

The nucleophilic attack by amines results in the displacement of chloro ligands as in

$$K_2[PtCl_3(C_2H_4)] + RNH_2 \rightarrow K_2[PtCl_2RNH_2(C_2H_4)]$$

The Zeise's salt finds an application in the test for hepatitis.

8.3 Ferrocene

Ferrocene is an organometallic compound of the general class metallocene with the molecular formula $Fe(\eta^5\text{-}C_5H_5)_2$. In this molecule, iron is sandwiched between two cyclopentadienyl rings in staggered conformation, as shown in Figure 6.

(a)   (b)

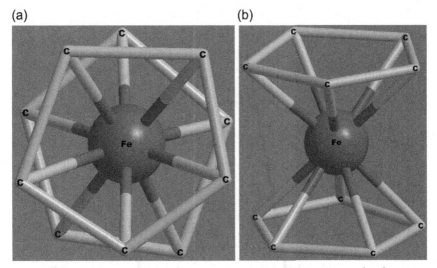

Figure 6 (a) Top and (b) side view of a staggered ferrocene molecule.

The five molecular orbitals of the two cyclopentadienyl ligands combine to give 10 ligand molecular orbitals in three energy levels, shown in the left part of Figure 7. These orbitals interact with the suitable metal orbitals to give a molecular orbital diagram for the ferrocene and related complexes. In the case of ferrocene, 10 electrons are contributed by the two cyclopentadiene molecules, whereas the Fe contributes 8 electrons, making it an 18-electron system. These 18 electrons are accommodated in the low lying molecular orbitals, excluding the antibonding ones. Thus, the stability of the ferrocene molecule is explained on the basis of molecular orbital theory.

Ferrocene is an orange-coloured diamagnetic solid. It is stable in air and sublimes at temperatures above $100\,°C$. The Fe atom of ferrocene readily oxidizes to Fe^{2+}, giving $Fe(C_5H_5)_2^{2+}$ ion.

The C—C bonds in ferrocene possess similar properties as C—C bonds in benzene. Hence, the reactions shown by benzene, due to aromatic character, are also exhibited by ferrocene. Ferrocene is represented in the following reactions as $Fe(\eta^5\text{-}C_5H_5)_2/Fe(Cp)_2$.

$$Fe(\eta^5-C_5H_5)_2 \xrightarrow[0\,°C]{CH_3COCl/AlCl_3} Fe(\eta^5-C_5H_4COCH_3)(\eta^5-C_5H_5)$$

$$\rightarrow Fe(\eta^5-C_5H_4COCH_3)_2 \text{ (Aromatic Acylation)}$$

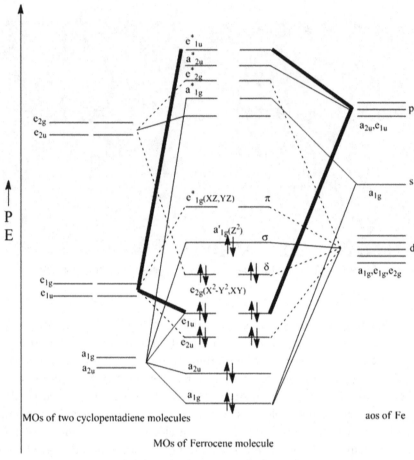

Figure 7 Molecular orbital diagram of a ferrocene molecule.

$$Fe(\eta^5\text{-}C_5H_5)_2 + nBuLi \rightarrow Fe(\eta^5\text{-}C_5H_4Li)(\eta^5\text{-}C_5H_5) + Fe(\eta^5\text{-}C_5H_4Li)_2$$
(Metallation)

$$Fe(\eta^5-C_5H_5)_2 \xrightarrow{\text{Li}/C_2H_5NH_2} Fe + 2C_5H_5^- \quad (\text{Reduction})$$

Ferrocene has found two interesting applications related to fuel. It is known to promote the smoke-free combustion of fuel. It also prevents knocking in the engine when used as an additive to a fuel. Ferrocene and some of its derivatives are also useful as curing accelerators for unsaturated polyester resins [5].

9. EXERCISES

9.1 Multiple Choice Questions

1. Which of the following elements is present in Cacodyl?
   (a) mercury   (b) arsenic
   (c) gold   (d) silver

2. The organometallic compounds of type MRn of main group IV to VI are _____.
   (a) not easily separated as adducts   (b) having very high Lewis
       with Lewis bases   acidity
   (c) strongly ionic in nature   (d) all of these

3. $MX_n + M'R_m \rightarrow MX_{n-a} R_a + M'R_{m-a} X_a$ (X = halogen and R = alkyl or aryl) is a general reaction for _____.
   (a) carbon—halide exchange   (b) metal—metal exchange
   (c) oxidative addition   (d) reductive elimination

4. Organoalkali compounds are the sources of _____.
   (a) free radicals   (b) electrophiles
   (c) powerful carbanions   (d) carbocations

9.2 Short/Long Answer Questions

1. Define an organometallic compound and give an example of a dihapto organometallic compound.
2. "$Ti(OC_4H_9)_4$ is not an organometallic compound but $C_6H_5Ti(OC_4H_9)_3$ is an organometallic compound". Why?
3. How is a Wurtz—Fittig reaction minimized in the preparation of organometallic compounds by oxidative—addition reactions?
4. Distinguish the ionic organometallic compounds from the σ-bonded organometallic compounds.
5. What is the role of the pK_a value of R—H in determining the stability of ionic organometallic compounds?
6. Give an account of fluxional organometallic compounds.
7. Discuss the structure and bonding in Zeise's salt.
8. Using molecular orbital theory, justify the stability of ferrocene.

SUGGESTED FURTHER READINGS

The topics discussed in this chapter are a part of a standard graduate curriculum. A majority of the textbooks with titles related to organometallic chemistry can act as a source of further reading. Moreover, there are several web resources useful for further learning. Some of them are listed below:

http://www.nptel.ac.in/courses/104101006/downloads/lecture-notes/mod2/lec4.pdf
https://en.wikipedia.org/wiki/Fluxional_molecule
http://www.nptel.ac.in/courses/104101006/downloads/lecture-notes/mod9/lec3.pdf
http://www.chem.ucalgary.ca/courses/350/Carey5th/Ch14/ch14-0.html
http://nptel.ac.in/courses/104106064/lectures.pdf
http://pac.iupac.org/publications/pac/pdf/1999/pdf/7108x1557.pdf

REFERENCES

Some illustrations are cited from the books and research articles listed below:
[1] Mehrotra RC. Organometallic chemistry. New Age International (P) Limited; 2007.
[2] Cotton FA. 10-Stereochemical nonrigidity in organometallic compounds. In: Cotton LMJA, editor. Dynamic nuclear magnetic resonance spectroscopy. Academic Press; 1975. p. 377–440.
[3] Mann BE. 20-Non-rigidity in organometallic compounds. In: Abel GWGASW, editor. Comprehensive organometallic chemistry. Oxford: Pergamon; 1982. p. 89–171.
[4] Shoemaker KA, Leadbeater NE. A fast and easy approach to the synthesis of Zeise's salt using microwave heating. Inorg Chem Commun 2009;12(5):341–2.
[5] Kalenda P. Ferrocene and some of its derivatives used as accelerators of curing reactions in unsaturated polyester resins. Eur Polym J 1995;31(11):1099–102.

CHAPTER 8

Metal Carbonyls

Contents

Essentials of Coordination Chemistry
http://dx.doi.org/10.1016/B978-0-12-803895-6.00008-2

1. INTRODUCTION

The history of metal carbonyls goes back to 1834 when Justus von Liebig attempted initial experiments on the reaction of carbon monoxide with metals. However, it was demonstrated later that the compound he claimed to be potassium carbonyl was not a metal carbonyl at all. After the synthesis of [PtCl$_2$(CO)$_2$] and [PtCl$_2$(CO)]$_2$ was reported by Schutzenberger (1868), followed by [Ni(CO)$_4$], reported by Mond (1890), Hieber prepared numerous compounds containing metal and carbon monoxide [1].

Compounds with at least one bond between carbon and metal are known as organometallic compounds [2]. Very strictly speaking, the carbon in organometallic compounds should be organic. Metal carbonyls are the transition metal complexes of carbon monoxide, containing a metal–carbon bond. These metal–carbon bonded complexes are such important players of organometallic chemistry that it cannot afford to keep the metal carbonyls out of the team just because of the definition. The metal carbonyls offer a very facile route to the synthesis of many other organometallic compounds.

A lone pair of electrons are available on both carbon and oxygen atoms of a carbon monoxide ligand. However, as the carbon atoms donate electrons to the metal, these complexes are named carbonyls. A variety of such complexes, such as mononuclear, polynuclear, homoleptic and mixed ligands, are known. These compounds are widely studied due to their ability to release carbon monoxide [3], their industrial importance, their catalytic properties [4] and their structural interest [5]. Carbon monoxide is one of the most important π-acceptor ligands. Because of its π-acidity, carbon monoxide can stabilize the zero formal oxidation state of metals in carbonyl complexes.

2. SYNTHESIS OF METAL CARBONYLS

The following are some of the general methods of preparation of metal carbonyls.

2.1 Direct Combination

Only Ni(CO)$_4$, Fe(CO)$_5$ and Co$_2$(CO)$_8$ are normally obtained by the action of carbon monoxide on a finely divided metal at a suitable temperature and pressure.

$$Ni(s) + 4CO(g) \xrightarrow{30\ °C,\ 1\ atm} Ni(CO)_4(l)$$

$$Fe(s) + 5CO(g) \xrightarrow{200\ °C,\ 200\ atm} Fe(CO)_5(l)$$

$$2Co(s) + 8CO(g) \xrightarrow{150\ °C,\ 35\ atm} Co_2(CO)_8(s)$$

2.2 Reductive Carbonylation

Many metallic carbonyls are obtained when salts like $Ru(acac)_3$, $CrCl_3$, Re_2O_7, VCl_3, CoS, $Co(CO)_3$ and CoI_2 are treated with carbon monoxide in the presence of a suitable reducing agent such as Mg, Ag, Cu, Na, H_2, $AlLiH_4$, etc.

$$3Ru(acac)_3(solution) + H_2(g)$$
$$+ 12CO(g) \xrightarrow{150\ °C,\ 200\ atm,\ methanol} Ru_3(CO)_{12}$$

$$CrCl_3(s) + Al(s) + 6CO(g) \xrightarrow{AlCl_3,\ benzene} Cr(CO)_6(solution)$$

$$2MnI_2 + 10CO + 2Mg \xrightarrow[ether]{25\ °C,\ 210\ atm} Mn_2(CO)_{10} + 2MgI_2$$

$$2CoS + 8CO + 4Cu \xrightarrow{200\ °C,\ 200\ atm\ press} Co_2(CO)_8 + 2Cu_2S$$

$$2CoI_2 + 8CO + 4Cu \xrightarrow{200°C,\ 200\ atm\ press} Co_2(CO)_8 + 4CuI$$

$$2FeI_2 + 5CO + 2Cu \xrightarrow{200\ °C,\ 200\ atm\ press} Fe(CO)_5 + Cu_2I_2$$

$$2CoCO_3 + 8CO + 2H_2 \xrightarrow{120-200\ °C,\ 250-300\ atm\ press} Co_2CO_8 + 2CO_2 + 2H_2O$$

$$MoCl_5 + 6CO + 5Na \xrightarrow{diglyme} Mo(CO)_6 + 5NaCl$$

Sometimes CO acts as both a carbonylating and reducing agent, as shown in the following reactions:

$$OsO_5 + 5CO \xrightarrow{250\ °C,\ 350\ atm\ press} Os(CO)_5 + 2O_2$$

$$Re_2O_7(s) + 17CO(g) \xrightarrow{250\ °C,\ 350\ atm} Re_2(CO)_{10}(s) + 7CO_2(g)$$

A solution of vanadium chloride in diethylene glycol dimethyl ether, which is acidified by phosphoric acid, gives vanadium hexacarbonyl as

$$VCl_3 + 6CO + 4Na + 2CH_3O(CH_2CH_2O)_2CH_3 \rightarrow$$
$$\{Na[CH_3O(CH_2CH_2O)_2CH_3]_2\}^+ [V(CO)_6]^- + 3NaCl$$
$$\{Na[CH_3O(CH_2CH_2O)_2CH_3]_2\}^+ [V(CO)_6]^- \xrightarrow{H^+} [V(CO)_6]$$

2.3 Preparation of Mononuclear Carbonyls from Iron Pentacarbonyl

The labile carbonyl groups in iron pentacarbonyl can be replaced by chloride to give a different metal carbonyl. These reactions are characterized by low yield, which can be improved using high pressure.

$$\text{MoCl}_6 + 3\text{Fe(CO)}_5 \xrightarrow{110\ °C,\ \text{ether}} \text{Mo(CO)}_6 + 3\text{FeCl}_2 + 9\text{CO}$$

$$\text{WCl}_6 + 3\text{Fe(CO)}_5 \xrightarrow{110\ °C,\ \text{ether}} \text{W(CO)}_6 + 3\text{FeCl}_2 + 9\text{CO}$$

2.4 Preparation of Dinuclear Carbonyls from Mononuclear Carbonyls

When a cold solution of $\text{Fe(CO)}_5/\text{Os(CO)}_5$ in glacial CH_3COOH is irradiated with ultraviolet light, $\text{Fe}_2(\text{CO})_9/\text{Os}_2(\text{CO})_9$ are obtained as

$$\text{Fe(CO)}_5 \xrightarrow{h\nu} \text{Fe}_2(\text{CO})_9 + \text{CO}$$

$$\text{Os(CO)}_5 \xrightarrow{h\nu} \text{Os}_2(\text{CO})_9 + \text{CO}$$

2.5 Preparation of Mixed-Metal Carbonyls by Metathesis Reaction

The following reaction between cobalt and ruthenium carbonyl uses metathesis to obtain a mixed metal complex:

$$\text{KCo(CO)}_4 + \left[\text{Ru(CO)}_3\text{Cl}_2\right]_2 \rightarrow 2\text{RuCo}_2(\text{CO})_{11} + 4\text{KCl}$$

3. PHYSICAL PROPERTIES

A majority of the metallic carbonyls are liquids or volatile solids. Most of the mononuclear carbonyls are colourless to pale yellow. V(CO)_6 is a bluish-black solid. Polynuclear carbonyls are dark in colour. Metal carbonyls are soluble in organic solvents like glacial acetic acid, acetone, benzene, carbon tetrachloride and ether. Due to low melting points and poor thermal stability, they show toxicity related to the corresponding metal and carbon monoxide. Exposure to these compounds can cause damage to the lungs, liver, brain and kidneys. Nickel tetracarbonyl exhibits the strongest inhalation toxicity. These compounds are carcinogenic over long-term exposure. All of the metal carbonyls, other than vanadium hexacarbonyl, are diamagnetic. The metals with an even atomic number form mononuclear carbonyls. Thus, all the electrons in the metal atoms are paired. In the case of dinuclear metal carbonyls formed by metals with an odd atomic number, the unpaired electrons are utilized for the formation of metal–metal bonds. Most of the metal carbonyls melt or decompose at low temperatures. Solid carbonyls sublime in a vacuum, but they undergo some degree of degradation. Metal carbonyls are thermodynamically unstable.

They undergo aerial oxidation with different rates. $Co_2(CO)_8$ and $Fe_2(CO)_9$ are oxidized by air at room temperature, while chromium and molybdenum hexacarbonyls are oxidized in air when heated.

4. CHEMICAL PROPERTIES

The metal carbonyls give a variety of chemical reactions.

4.1 Ligand Substitution Reactions

The substitution of a carbon monoxide ligand by various monodentate and bidentate ligands can be carried out using thermal and photochemical reactions. Monodentate ligands such as isocyanides (CNR), cyanide (CN^-), phosphine (PR_3) and ethers can partially or completely replace the carbonyl group.

$$Fe(CO)_5 + 2CNR \rightarrow Fe(CO)_3(CNR)_2 + 2CO$$
$$Ni(CO)_4 + 4CNR \rightarrow Ni(CNR)_4 + 4CO$$
$$Mn_2(CO)_{10} + PR_3 \rightarrow 2Mn(CO)_4(PR_3) + 2CO$$
$$2Fe_2(CO)_{12} + 3py \rightarrow Fe_3(CO)_9(py)_3 + 3Fe(CO)_5$$

Bidentate ligands such as o-phenylene–bis(dimethyl arsine) (diars) and o-phenanthroline(o-phen) can replace carbonyl groups in the multiple of two.

$$Mo(CO)_6 + diars \rightarrow Mo(CO)_4(diars) + 2CO$$
$$Ni(CO)_4 + o\text{-phen} \rightarrow Ni(CO)_2(o\text{-phen})_2 + 2CO$$
$$Cr(CO)_6 + 2diars \rightarrow Cr(CO)_2(diars)_2 + 4CO$$

4.2 Reaction with Metallic Sodium

Metallic sodium and its amalgam can be used to reduce the metal carbonyls.

$$Cr(CO)_6 + 2Na \rightarrow Na_2[Cr(CO)_5] + CO$$
$$Mn_2(CO)_{10} + 2Na \rightarrow 2Na[Mn(CO)_5] + CO$$

In these two reactions, the Cr and Mn atoms in their zero oxidation states are reduced to -2 and -1 oxidation states, respectively.

4.3 Reaction with Sodium Hydroxide

The reaction of sodium hydroxide with metal carbonyls results in a nucleophilic attack by a hydroxide ion on the carbonyl group to give a metal carboxylic acid complex. Upon further action with sodium hydroxide, the carboxylic acid gives up carbon dioxide to form a hydrido anion. The protonation of this anion results in the formation of iron tetracarbonyl hydride as

$$Fe(CO)_5 + NaOH \rightarrow Na[Fe(CO)_4COOH]$$
$$Na[Fe(CO)_4COOH] + NaOH \rightarrow Na[HFe(CO)_4] + NaHCO_3$$
$$Na[HFe(CO)_4] + H^+ \rightarrow (H)_2Fe(CO)_4 + Na^+$$

This reaction is known as a Heiber base reaction.

4.4 Reaction with Halogens

Most of the metal carbonyls react with halogens to give carbonyl halides

$$Fe(CO)_5 + X_2 \rightarrow Fe(CO)_4X_2 + CO$$
$$Mo(CO)_6 + Cl_2 \rightarrow Mo(CO)_4Cl_2 + 2CO$$

Halogens can cause cleavage in the metal–metal bonds in the case of polynuclear carbonyls.

$$Mn_2(CO)_{10} + X_2 \rightarrow 2Mn(CO)_5X$$

Some carbonyls undergo decomposition upon reaction with halogens.

$$Ni(CO)_4 + Br_2 \rightarrow NiBr_2 + 4CO$$
$$Co_2(CO)_8 + 2X_2 \rightarrow 2CoX_2 + 8CO$$

4.5 Reaction with Hydrogen

Some of the carbonyls can be reduced by hydrogen to give carbonyl hydrides.

$$Co_2(CO)_8 + H_2 \xrightarrow{165\ °C,\ 200\ atm} 2[Co(CO)_4H]$$
$$Mn_2(CO)_{10} + H_2 \xrightarrow{200\ atm} 2[Mn(CO)_5H]$$

Even though these compounds are named as hydrides, they are known to behave as proton donors. The neutral hydrides, such as [Co(CO)$_4$H] and [Mn(CO)$_5$H], behave as acids as

$$[Co(CO)_4H] \rightarrow [Co(CO)_4]^- + H^+$$
$$[Mn(CO)_5H] \rightarrow [Mn(CO)_5]^- + H^+$$

The anionic hydrides, such as $[HFe(CO)_4]^-$, are true hydrides and behave as reducing agents for alkyl halides as

$$RX + [HFe(CO)_4]^- \rightarrow RH + [XFe(CO)_4]^-$$

4.6 Reaction with Nitric Oxide

A good number of metal carbonyls react with nitric oxide to give carbonyl nitrosyls.

$$Fe(CO)_5 + 2NO \xrightarrow{\text{95 °C}} Fe(CO)_2(NO)_2 + 3CO$$
$$Co_2(CO)_8 + 2NO \xrightarrow{\text{40 °C}} 2Co(CO)_3(NO) + 2CO$$

The reaction between iron pentacarbonyl and nitric oxide involves the replacement of three carbonyl groups by two nitric oxide molecules. Electronically, this is the equivalent, as nitric oxide is a three-electron donor ligand, whereas carbon monoxide is a two-electron donor.

5. BONDING IN METALLIC CARBONYLS

5.1 Carbon Monoxide

In order to understand the bonding in metal carbonyls, let us first see the molecular orbital (MO) diagram of carbon monoxide, shown in Figure 1.

The order of energy of the molecular orbitals and the accommodation of 10 electrons of the carbon monoxide can be shown as $(\sigma_s^b)^2(\sigma_p^b)^2$ $(\pi_y^b = \pi_z^b)^4(\sigma_s^*)^2(\pi_y^* = \pi_z^*)^0(\sigma_p^*)^0$

From the figure, it is seen that (σ_s^*) is the highest occupied molecular orbital (HOMO) that can donate the lone pair of electrons for the formation of an $OC \rightarrow M$ σ bond.

While $(\pi_y^* = \pi_z^*)$ are the lowest unoccupied molecular orbitals (LUMO) that can accept the electron density from an appropriately oriented filled metal orbital, resulting into the formation of an $M \rightarrow CO$ π bond.

Figures 2 and 3 show the highest occupied molecular orbital (HOMO) and lowest unoccupied molecular orbital (LUMO) for the carbon monoxide molecule. In the figures, white colour is for the positive sign of the wave function, while black colour indicates the negative sign of the wave function.

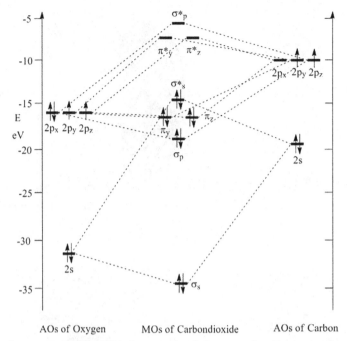

AOs of Oxygen   MOs of Carbondioxide   AOs of Carbon

Figure 1 Molecular orbital energy level diagram of carbon monoxide.

The nature of M—CO bonding in mononuclear carbonyls can be understood by considering the formation of a dative σ-bond and π-bond due to back-donation.

5.2 Formation of Dative σ-Bond

The overlapping of an empty hybrid orbital, which is a blend of d-, s- and p-orbitals on a metal atom with the filled hybrid orbital (HOMO) on a carbon atom of carbon monoxide molecule, results into the formation of an M←CO σ-bond, as shown in Figure 4.

5.3 Formation of π-Bond by Back-Donation

This bond is formed because of the overlapping of filled dπ orbitals or hybrid dpπ orbitals of a metal atom with low-lying empty (LUMO) orbitals on a CO molecule, i.e. M $\overset{\pi}{\rightarrow}$ CO, as shown in Figure 5.

5.4 Bridging CO Groups

In addition to the linear M—C—O groups, the carbon monoxide ligand is also known to form bridges. This type of bonding is observed in some

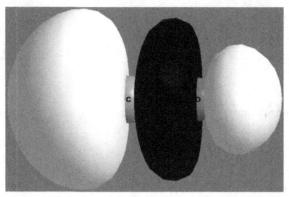

Figure 2 Highest occupied molecular orbital (HOMO) of carbon monoxide.

Figure 3 Lowest unoccupied molecular orbital (LUMO) of carbon monoxide.

Figure 4 Formation of an M←CO σ-bond in metal carbonyls.

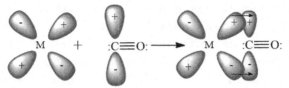

Figure 5 Formation of M $\xrightarrow{\pi}$ CO bond by back-donation in metal carbonyls.

binuclear and polynuclear carbonyls. It is denoted by μ_n—CO, where n indicates the number of metals bridged. While $n = 2$ is the most common value, it reaches 3 or 4 in certain, less common carbonyls. In a terminal M—C—O group, the carbon monoxide donates two electrons to an empty metal orbital, while in μ_2—CO group, the M—C bond is formed by the sharing of one metal electron and one carbon electron.

6. INFRARED SPECTROSCOPY

The carbonyl groups can have two modes of stretching, as shown in Figure 6.

Since both of these modes result in change in dipole moment, two bands are expected in the infrared spectra of a terminally ligated carbon monoxide. The infrared and Raman spectroscopy together can be used to determine the geometry of the metallic carbonyls. A mononuclear penta-carbonyl can exist both in square pyramidal and trigonal bipyramidal geometry. Performing infrared spectra after calculating the Infrared (IR) active and Raman active bands in both the possible geometries can provide information about the actual geometry of the molecule. Infrared

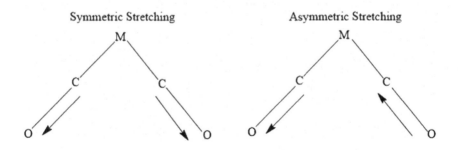

Figure 6 Stretching modes of carbonyl.

spectroscopy of metallic carbonyls also helps in determining the bond order of ligated carbon monoxide. The C—O bond order and the frequency related to its absorption are directly proportional. Thus, it can be predicted that the frequencies of absorption will be in the following order: Free CO > metal carbonyl cation > neutral metal carbonyl > metal carbonyl anion, which is summarized in Table 1.

It is also used to distinguish the terminal and bridging carbonyl groups, as shown in Figure 7.

The C—O bonding in terminal carbonyl groups is stronger than the bridged carbonyl groups. Therefore, it is possible to differentiate the terminal carbonyls, which absorb in the region of $2050–1900$ cm^{-1}, from the bridged carbonyls, absorbing below 1900 cm^{-1}. The change in the intensity of bands related to the carbonyl group can provide information for the kinetic studies of the substitution reactions involving the replacement of carbonyls.

Table 1 Comparison C—O stretching in representative metal carbonyls

Carbonyl	Type	C—O stretching frequency (cm^{-1})
Carbon monoxide	Free	~ 2150
$Mn(CO)_6^+$	Cation	~ 2090
$Cr(CO)_6$	Neutral	~ 2000
$V(CO)_6^-$	Anion	~ 1850

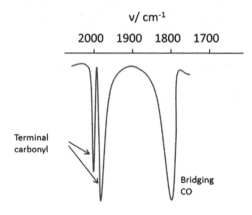

v for free CO = ~2150 cm^{-1}

Figure 7 A partial infrared spectrum showing terminal and bridged carbonyl.

7. CLASSIFICATION OF METAL CARBONYLS

The metal carbonyls have been an area of research interest to many, and hence, a wide variety of such compounds have been prepared and characterized. For the scope of the present chapter, a classification of metal carbonyl is provided with some examples in Table 2.

8. MONONUCLEAR CARBONYLS

8.1 Ni(CO)$_4$, Nickel Tetracarbonyl

8.1.1 Preparation

It can be prepared by passing carbon monoxide over nickel in the temperature range of 60–100 °C.

$$Ni + 4CO \xrightarrow{60\,°C} Ni(CO)_4$$

It can be made by heating nickel iodide with carbon monoxide in the presence of copper, which acts as a halogen acceptor.

$$NiI_2 + 4CO \xrightarrow{Cu} Ni(CO)_4 + CuI_2$$

It can also be prepared by passing carbon monoxide through alkaline suspensions of nickel sulphide or nickel cyanate.

$$NiS + 4CO \rightarrow Ni(CO)_4 + S$$
$$Ni(CN)_2 + 4CO \rightarrow Ni(CO)_4 + C_2N_2$$

8.1.2 Properties

It is a colourless liquid with a melting point of −25 °C, a boiling point of 43 °C and a decomposition temperature in the range of 180–200 °C.

Table 2 Some varieties of metal carbonyls with examples

Type	Examples
Mononuclear carbonyls	$[Ti(CO)_6]^{-2}$, $[V(CO)_6]$, $[Cr(CO)_6]$, $[Fe(CO)_5]$, $[Ni(CO)_4]$
Dinuclear carbonyls	$[Mn_2(CO)_{10}]$, $[Fe_2(CO)_9]$, $[Co_2(CO)_8]$
Polynuclear carbonyls	$[Fe_3(CO)_{12}]$, $[Co_4(CO)_{12}]$, $[Co_6(CO)_{16}]$
μ_2-bridging carbonyls	$[Fe_2(CO)_9]$, $[Co_2(CO)_8]$, $[Fe_3(CO)_{12}]$, $[Co_4(CO)_{12}]$
μ_3-bridging carbonyls	$[Rh_6(CO)_{16}]$ (four triply bridged carbonyl groups)
Carbonyl hydrides	$[HMn(CO)_5]$, $[HCo(CO)_4]$, $[H_2Fe(CO)_4]$

It is insoluble in water but dissolves in organic solvents.

It reacts with concentrated sulphuric acid along with detonation.

$$Ni(CO)_4 + H_2SO_4 \rightarrow NiSO_4 + H_2 + 4CO$$

It reacts with moist nitric oxide to give a deep blue-coloured compound.

$$2Ni(CO)_4 + 2NO + 2H_2O \rightarrow 2Ni(NO)(OH) + 8OH^- + H_2$$

Passing gaseous hydrochloric acid in the solution of nickel tetracarbonyl results in the decomposition

$$Ni(CO)_4 + 2HCl(g) \rightarrow NiCl_2 + H_2 + 4CO$$

8.1.3 Uses

Since $Ni(CO)_4$, on heating, decomposes to metallic nickel, it is used in the production of nickel by Mond's process.

It is used for plating nickel on other metals.

It is used as a catalyst for the synthesis of acrylic monomers in plastic industries.

8.1.4 Structure

Nickel tetracarbonyl has a tetrahedral geometry with Ni—C bond lengths of 1.5 Å. It is also found to be diamagnetic.

The structure of $Ni(CO)_4$, shown in Figure 8, can be explained by considering sp^3 hybridization of Ni atom. Since it is diamagnetic, all of the 10 electrons present in the valence shell of Ni atom $(Ni = 3d^8\ 4s^2)$ get paired in three 3d orbitals. Thus, the valence shell configuration of Ni atom in $Ni(CO)_4$ molecule becomes $3d^{10}\ 4s^0$. $OC \rightarrow Ni$ bond results by the overlap between the empty sp^3 hybrid orbital on Ni atom and the HOMO on C atom in CO molecule, as shown in Figure 9.

Acceptance of four electron pairs by nickel in a zero oxidation state severely increases the electron density on the nickel atom. According to the electro neutrality principle given by Pauling, the atoms in a molecule share the electron pairs to the extent that the charge on each of the atoms remains close to zero. Thus, the nickel atom donates back some electron density from the filled d-orbitals to the low-lying empty (LUMO) orbitals on CO molecule, resulting into the formation of a double bond, i.e. $M \xrightarrow{\pi} CO$.

(a)

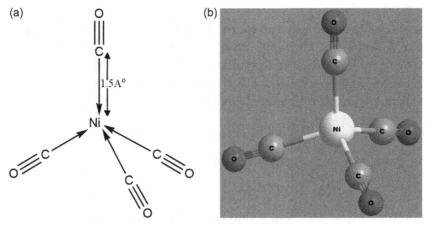

(b)

Figure 8 Tetrahedral structure of nickel tetracarbonyl.

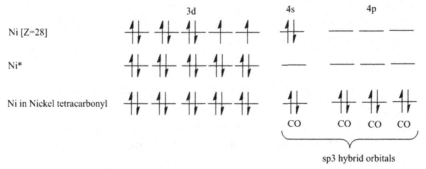

Figure 9 sp³ hybridization of a nickel atom in nickel tetracarbonyl.

8.2 Fe(CO)₅, Iron Pentacarbonyl

8.2.1 Preparation

It can be prepared by passing carbon monoxide over iron powder at a high temperature and pressure.

$$Fe + 5CO \xrightarrow{\text{200 °C, 100 atm}} Fe(CO)_5$$

It can also be prepared by the carbonylation of ferrous sulphide/iodide in the presence of Cu-metal, which acts as a reducing agent.

$$2FeS + 10CO + 2Cu \xrightarrow{\text{200 °C, 200 atm}} Fe(CO)_5 + Cu_2S$$

8.2.2 Properties

It is a pale-yellow liquid with a melting point of $-20\,°C$, a boiling point of $103\,°C$ and a decomposition temperature around $250\,°C$. It is insoluble in water but soluble in glacial acetic acid, methanol, diethyl ether, acetone and benzene. A cold solution of iron pentacarbonyl in glacial acetic acid undergoes dimerization under the influence of ultraviolet light.

$$2Fe(CO)_5 \xrightarrow{h\nu} Fe_2(CO)_9 + CO$$

It is readily hydrolysed by water and acids.

$$Fe(CO)_5 + H_2SO_4 \rightarrow FeSO_4 + 5CO + H_2$$

The reaction of sodium hydroxide with iron pentacarbonyl results in a nucleophilic attack by a hydroxide ion on the carbonyl group to give a metal carboxylic acid complex. Upon further action with sodium hydroxide, the carboxylic acid gives up carbon dioxide to form a hydrido anion. The protonation of this anion results in the formation of iron tetracarbonyl hydride (Heiber base) as

$$Fe(CO)_5 + NaOH \rightarrow Na[Fe(CO)_4COOH]$$
$$Na[Fe(CO)_4COOH] + NaOH \rightarrow Na[HFe(CO)_4] + NaHCO_3$$
$$Na[HFe(CO)_4] + H^+ \rightarrow [(H)_2Fe(CO)_4] + Na^+$$

It reacts with sodium metal in liquid ammonia to give a carbonylate anion.

$$Fe(CO)_5 + 2Na \xrightarrow{liquid\ NH_3} Na_2[Fe(CO)_4] + CO$$

This compound is popularly known as Collman's reagent in organic synthesis. The Collman's reagent is used in aldehyde synthesis as

$$Na_2[Fe(CO)_4] + RBr \xrightarrow{liquid\ NH_3} Na[RFe(CO)_4] + NaBr$$

This solution is then treated with triphenyl phosphine followed by acetic acid to give the corresponding aldehyde.

It reacts with ammonia to give iron tetracarbonyl hydride and carbamic acid (glycine).

$$Fe(CO)_5 + H_2O + NH_3 \rightarrow Fe(CO)_4(H)_2 + NH_2COOH$$

It reacts with halogens in nonaqueous solvents to give stable tetra-carbonyl halides.

$$Fe(CO)_5 + X_2 \rightarrow Fe(CO)_4X_2 + CO$$

8.2.3 Structure

The structural studies have suggested trigonal bipyramidal geometry for iron pentacarbonyl. The Fe—C distances are found to be 1.80 Å and 1.84 Å for axial and equatorial bonds, respectively. The molecule is also found to be diamagnetic.

The structure shown in Figure 10 can be explained using dsp^3 hybridization in Fe atom. All eight electrons present in the valence shell of Fe atom ($Fe:3d^64s^2$) get paired in four 3d orbitals. Thus, the valence shell configuration of Fe in $Fe(CO)_5$ becomes $3d^84s^0$. The $OC \rightarrow Fe$ bond results by the overlap between the empty dsp^3 hybrid orbitals on Fe atom and the HOMO on C atom in a CO molecule, as shown in Figure 11.

8.3 Cr(CO)₆, Chromium Hexacarbonyl

8.3.1 Preparation

It can be prepared by carbonylation of chromium chloride with carbon monoxide using a reducing agent such as lithium aluminium hydride (LAH).

$$CrCl_3 + 6CO \xrightarrow{\text{LAH, 115 °C, 70 atm}} Cr(CO)_6$$

An indirect method of preparation involves an action of carbon monoxide on a mixture of Grignard reagent and anhydrous chromium chloride

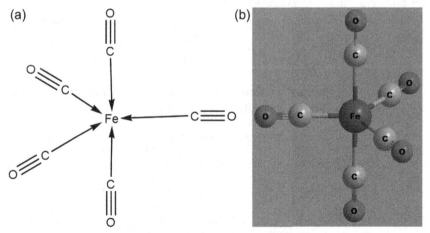

Figure 10 Trigonal bipyramidal structure of iron pentacarbonyl.

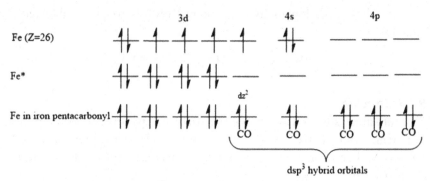

Figure 11 dsp³ hybridization in iron pentacarbonyl.

in ether, which is followed by decomposition with an acid to give chromium hexacarbonyl.

$$C_6H_5MgBr + CrCl_3 + CO \xrightarrow{35-70 \text{ atm}} Cr(CO)_2(C_6H_5)_4 + MgBrCl +$$

$$MgBr_2Cr(CO)_2(C_6H_5)_4 + 6H^+ \rightarrow Cr(CO)_6 + 2Cr^{+3} + 12(C_6H_5)^- + 3H_2$$

8.3.2 Properties

It is a white crystalline solid, melting above 150 °C and boiling at 220 °C.

It is insoluble in water but soluble in ether, chloroform, carbon tetrachloride and benzene.

It is not attacked by air, bromine, cold aqueous alkalis, dilute acids, concentrated hydrochloric acid and sulphuric acid. It is decomposed by chlorine gas and concentrated nitric acid. It reacts with fluorine at −75 °C to form chromium hexafluoride.

It reacts with sodium metal in liquid ammonia to give a carbonylate anion.

$$Cr(CO)_6 + 2Na \xrightarrow{\text{liquid NH}_3} Na_2^+\left[Cr^{2-}(CO)_5\right]^{2-} + CO$$

It gives substitution reactions with amines like en and py. At higher temperatures (>150 °C), several pyridyl derivatives are formed.

$$Cr(CO)_6 + 2py \rightarrow \underset{\text{yellowish brown}}{Cr(CO)_4(py)_2} + 2CO$$

$$2Cr(CO)_6 + 5py \rightarrow \underset{\text{orange}}{Cr_2(CO)_7(py)_5} + 5CO$$

$$Cr(CO)_6 + 3py \rightarrow \underset{\text{bright red}}{Cr(CO)_3(py)_3} + 3CO$$

8.3.3 Structure

The structural studies have suggested an octahedral geometry for chromium hexacarbonyl. The Cr—C distance is found to be 1.92 Å, while the C—O bond length is 1.16 Å. The molecule is also found to be diamagnetic.

The structure shown in Figure 12 can be explained using d^2sp^3 hybridization in Cr atom. All six electrons present in the valence shell of Cr atom (Cr:$3d^54s^1$) get paired in three 3d orbitals. Thus, the valence shell configuration of Cr in $Cr(CO)_6$ becomes $3d^64s^0$. The OC→Cr bond results by the overlap between the empty d^2sp^3 hybrid orbitals on Fe atom and the HOMO on C atom in a CO molecule, as shown in Figure 13.

The MO energy diagram for $Cr(CO)_6$ is shown in Figure 14. For the molecular orbitals, 12 electrons are contributed from the lone pairs on the carbon atoms of the six-carbon monoxide ligands. The metal contributes six electrons, while 24 electrons come from the π-system of the six ligands. The MOs are occupied by these 42 electrons, and the t_{2g} level becomes the HOMO of the metal carbonyl.

The net effect of the π* orbitals is to increase the magnitude of 10 Dq (the splitting between the t_{2g} and e_g levels by lowering t^*_{2g} to a level lower in energy than when no π* orbitals are involved). Consequently, the complexes are predicted to be more stable when the ligands have π and π* orbitals available for bonding. The ligand CO may be predicted to bond increasingly strongly with electron-releasing metal atoms. The bond order of CO decreases progressively as the π* orbitals are increasingly populated by d→π* donation.

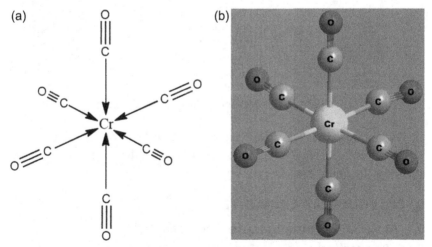

Figure 12 Octahedral structure of chromium hexacarbonyl.

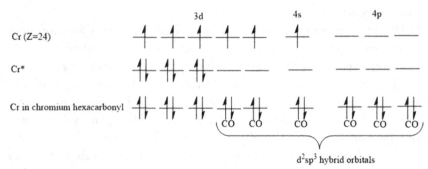

Figure 13 d^2sp^3 hybridization in chromium hexacarbonyl.

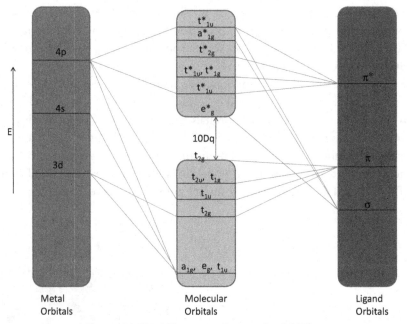

Figure 14 The MO energy diagram for $Cr(CO)_6$.

As discussed ahead, the low-lying empty π^* orbitals on CO allow back-bonding from the metal d-electrons to the ligand. It has a very pronounced effect on the coordinated C—O bond order.

9. POLYNUCLEAR CARBONYLS

9.1 $Mn_2(CO)_{10}$, Dimanganese Decacarbonyl

9.1.1 Preparation
It can be prepared by carbonylation of manganese iodide with carbon monoxide, using magnesium as a reducing agent.

$$2MnI_2 + 10CO + \underset{\text{(diethyl ether)}}{2Mg} \xrightarrow{25\ ^\circ C,\ 210\ atm} Mn_2(CO)_{10} + 2MgI_2$$

It can also be obtained by carbonylation of anhydrous manganese chloride with carbon monoxide in the presence of sodium benzophenone ketyl.

$$2MnCl_2 + 10CO + 4(C_6H_5)_2CONa \xrightarrow{165\ ^\circ C,\ 140\ atm} Mn_2(CO)_{10}$$
$$+ 4(C_6H_5)_2CO + 4NaCl$$

9.1.2 Properties

It forms stable, golden-yellow crystals with a melting point of 155 °C.

It is oxidized by a trace amount of oxygen in a solution. Hence, the solution must be stored in an inert atmosphere.

Halogenation of dimanganese decacarbonyl proceeds with the breaking of a Mn—Mn bond and the formation of carbonyl halides.

$$Mn_2(CO)_{10} + X_2(X = Br, I) \rightarrow 2Mn(CO)_5X$$

It reacts with sodium metal in liquid ammonia to give a carbonylate anion.

$$Mn_2(CO)_{10} + 2Na \xrightarrow{\text{liquid } NH_3} 2Na^+ \left[Mn(CO)_5\right]^-$$

9.1.3 Structure

Manganese pentacarbonyl does not exist, as Mn $(Z = 25)$ has an odd atomic number. However, the structure of dimanganese decacarbonyl consists of two manganese pentacarbonyl groups joined through a Mn—Mn (2.79 Å) bond, as shown in Figure 15. The formation of this intermetallic bond effectively adds one electron to each of the manganese atoms. Thus, manganese, an element with an odd atomic number, forms a binuclear carbonyl. Since the molecule does not have any unpaired electrons, it is diamagnetic. The remaining two members of group VIIB, viz, Technetium (Tc) and Rhenium (Re), also form decacarbonyls with similar structures.

The structure of dimanganese decacarbonyl can be explained using d^2sp^3 hybridization, as shown in Figure 16. Out of the six hybrid orbitals on each manganese atom, five orbitals accept a lone pair of electrons from the carbon monoxide molecules to form 10 Mn←CO coordinate bonds. While, the remaining one half-filled orbital on each manganese overlap to form a Mn—Mn bond.

(a) 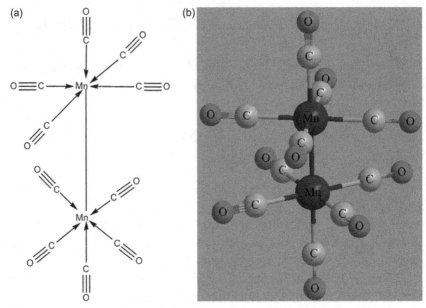 (b)

Figure 15 Structure of dimanganese decacarbonyl.

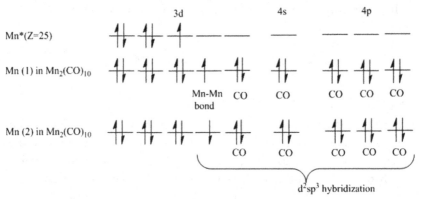

Figure 16 d^2sp^3 hybridization in dimanganese decacarbonyl.

9.2 $Co_2(CO)_8$, Dicobalt Octacarbonyl

9.2.1 Preparation

It can be prepared by direct combination of carbon monoxide with cobalt metal.

$$2Co + 8CO \xrightarrow{200\ °C,\ 100\ atm} Co_2(CO)_8$$

It can also be prepared by carbonylation of cobalt iodide/cobalt sulphide/cobalt carbonate, using reducing agents such as copper metal or hydrogen gas.

$$2CoS/2CuI_2 + 8CO + 4Cu \xrightarrow{200\ °C,\ 200\ atm} Co_2(CO)_8 + 2Cu_2S/4CuI$$

$$2CoCO_3 + 8CO + 2H_2 \xrightarrow{120-200\ °C,\ 250-300\ atm} Co_2(CO)_8 + 2H_2O$$

9.2.2 Properties

It is an orange crystalline substance with a melting point of 51 °C and turns deep violet upon exposure to air.

It is soluble in alcohols, ether and carbon tetrachloride.

Upon heating at 50 °C, it forms tetracobalt dodecacarbonyl.

$$2Co_2(CO)_8 \xrightarrow{50\ °C} Co_4(CO)_{12} + 4CO$$

It reacts with nitric oxide to form a cobalt carbonyl nitrosyl.

$$Co_2(CO)_8 + 2NO \rightarrow \left[Co^-(CO)_8(NO)^+\right]^0 + 2CO$$

9.2.3 Structure

Dicobalt octacarbonyl is known to exist in two isomeric forms, as shown in Figure 17. A bridged structure of this molecule is observed in the solid state, as well as a solution state at a very low temperature. A non-bridged structure predominates in a solution at temperatures above ambience.

In the bridged structure, the cobalt atoms are in a d^2sp^3 hybrid state, as shown in Figure 18. Three such hybrid orbitals on each cobalt atom accept a lone pair of electrons from three carbon monoxide molecules to form a total of six $Co \leftarrow CO$ coordinate bonds. A $Co-Co$ bond is formed by the overlapping of two half-filled d^2sp^3 hybrid orbitals on the cobalt atoms. The remaining two half-filled orbitals on each Co atom overlap with the appropriate orbital on a carbon atom of the carbonyl to form two bridging CO groups. Thus, all electrons in this molecule are paired and it is diamagnetic.

In the structure without the bridge, shown in Figure 19, the cobalt atoms are in dsp^3 hybrid state. Out of the five hybrid orbitals on each cobalt atom, four orbitals on each cobalt atom accept a lone pair of electrons from the carbon monoxide molecules to form eight $Co \leftarrow CO$ coordinate bonds. One half-filled orbital on each cobalt overlap to form a $Co-Co$ bond.

In the case of a non-bridge structure, Co atoms have dsp^3 hybridization.

(a)

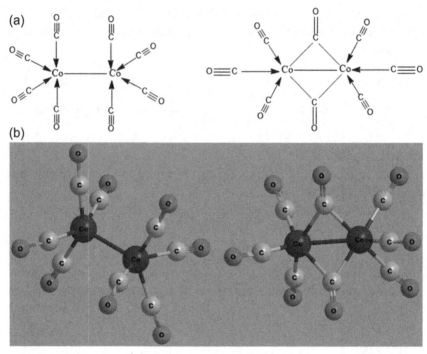

(b)

Figure 17 Structure of dicobalt octacarbonyl (without bridge and with bridge).

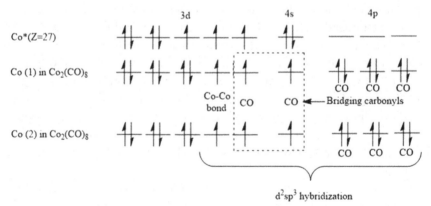

d²sp³ hybridization

Figure 18 d²sp³ hybridization in dicobalt octacarbonyl.

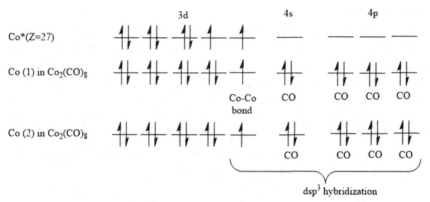

Figure 19 dsp^3 hybridization in dicobalt octacarbonyl.

9.3 Fe$_2$(CO)$_9$, Diiron Nonacarbonyl

9.3.1 Preparation

A cold solution of iron pentacarbonyl in glacial acetic acid undergoes dimerization under the influence of ultraviolet light to give golden–yellow crystals.

$$2Fe(CO)_5 \xrightarrow{hn} Fe_2(CO)_9 + CO$$

9.3.2 Properties

Diiron nonacarbonyl forms golden–yellow triclinic crystals, melting at 100 °C.

It is insoluble in water but soluble in toluene and pyridine.

A solution of diiron nonacarbonyl in toluene disproportionates when heated to 70 °C.

$$3Fe_2(CO)_9 \xrightarrow{70\ °C,\ Toluene} 3Fe(CO)_5 + Fe_3(CO)_{12}$$

Diiron nonacarbonyl reacts with sodium metal in liquid ammonia to give a carbonylate anion.

$$Fe_2(CO)_9 + 4Na \xrightarrow{liquid\ NH_3} Na_2{}^+ \left[Fe^{2-}(CO)_4\right]^{2-} + CO$$

9.3.3 Structure

Each of the iron atoms in diiron nonacarbonyl has three terminal carbonyl groups, as shown in Figure 20. The remaining three carbon monoxide ligands act as μ_2-CO groups. In addition to this, there is a weak Fe—Fe bond (2.46 Å) formed by the sharing of two unpaired electrons present in the 3d orbitals of iron atoms. Thus, both of the iron atoms in the molecule are identical with coordination number seven. Since the molecule does not have any unpaired electrons, it is diamagnetic.

The structure of this molecule can be explained using d^2sp^3 hybridization in Fe atoms, as shown in Figure 21.

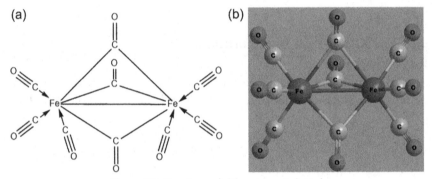

Figure 20 Structure of diiron nonacarbonyl.

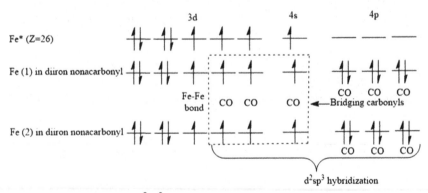

Figure 21 d^2sp^3 hybridization in diiron nonacarbonyl.

9.4 Fe₃(CO)₁₂, Triiron Dodecacarbonyl

9.4.1 Preparation

It is prepared by heating diiron nonacarbonyl dissolved in toluene at 70 °C.

$$3Fe_2(CO)_9 \xrightarrow{70\ °C,\ Toluene} 3Fe(CO)_5 + (CO)_{12}$$

It can also be prepared by the oxidation of iron carbonyl hydride.

$$3Fe(CO)_4H_2 + 3MnO_2 + 3H_2SO_4 \rightarrow Fe_3(CO)_{12} + 3MnSO_4 + 6H_2O$$

$$3Fe(CO)_4H_2 + 3H_2O_2 \rightarrow Fe_3(CO)_{12} + 6H_2O$$

9.4.2 Properties

It forms green monoclinic crystals, which are soluble in organic solvents like toluene, alcohol, etc. It decomposes at 140 °C to give metallic iron and carbon monoxide.

$$Fe_3(CO)_{12} \xrightarrow{140\ °C} 3Fe + 12CO$$

It gives a substitution reaction with pyridine and methanol.

$$3Fe_3(CO)_{12} + 3py \rightarrow Fe_3(CO)_9(py)_3 + 3Fe(CO)_5$$

$$2Fe_3(CO)_{12} + 3CH_3OH \rightarrow Fe_3(CO)_9(CH_3OH)_3 + 3Fe(CO)_5$$

It reacts with sodium metal in ammonia to give a carbonylate anion.

$$Fe_3(CO)_{12} + 6Na \xrightarrow{liquid\ NH_3} 3Na_2{}^+\left[Fe^{2-}(CO)_4\right]^{2-}$$

It reacts with nitric oxide to form iron dicarbonyl dinitrosyl.

$$Fe_3(CO)_{12} + 6NO \xrightarrow{85\ °C} 3Fe(CO)_2(NO)_2 + 6CO$$

9.4.3 Structure

Triiron dodecacarbonyl has a three-membered ring structure, as shown in Figure 22. Two iron atoms in this molecule have three terminal carbonyl groups, while the third iron atom is connected to four terminal carbonyls. Two μ₂−CO groups also connect the former iron atoms. In addition to this, there are three Fe−Fe bonds (2.8 Å) connecting each of the iron atoms.

(a)   (b)

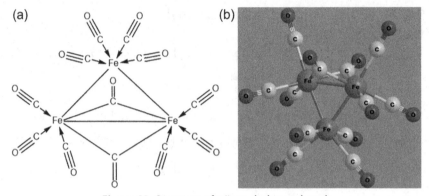

Figure 22 Structure of triiron dodecacarbonyl.

10. STRUCTURES OF A FEW POLYNUCLEAR CARBONYLS

In crystalline trinuclear dodecacarbonyls of ruthenium and osmium, the metal atoms form an equilateral triangle by metal–metal bonds. Each metal atom is bound to four terminal carbonyl groups, as shown in Figure 23. Two carbonyl groups are nearly perpendicular to (and the other two are parallel to) the plane of the triangle.

Figure 23 Structure of triosmium dodecacarbonyl.

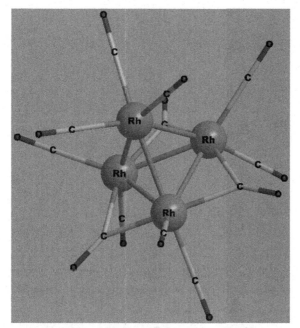

Figure 24 Structure of tetraosmium dodecacarbonyl.

The tetranuclear dodecacarbonyls of rhodium and cobalt have similar structures. Of the four Rh atoms arranged at the corners of a tetrahedron, one of the Rh atoms bears three terminal carbonyl groups. This atom does not have any bridging carbonyls. The remaining three Rh atoms, forming an equilateral triangle, bear two terminal carbonyls, and they are also bridged, as shown in Figure 24. All the rhodium atoms are connected by metal–metal bonds.

The structures of two more tetranuclear osmium carbonyls are shown in Figure 25 and Figure 26.

11. EFFECTIVE ATOMIC NUMBER (EAN) RULE

Effective Atomic Number (EAN) is the total number of electrons surrounding the nucleus of a metal in a complex.

Sidgwick's EAN rule/Inert gas rule:

"The EAN of the metal atom in a stable complex is equal to the atomic number of a noble gas found in the same period of the periodic table."

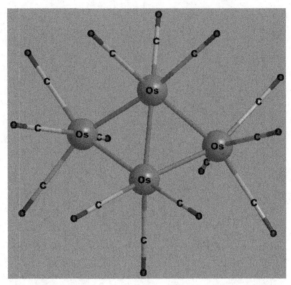

Figure 25 Structure of tetraosmium tetradecacarbonyl.

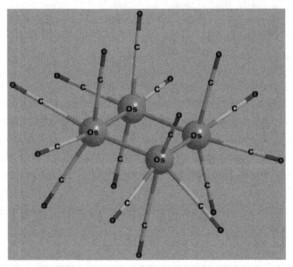

Figure 26 Structure of tetraosmium hexadecacarbonyl.

Most of the organometallic compounds, including carbonyls and nitrosyls, obey the EAN rule.

It is mainly useful in predicting the number of ligands attached to the metal in such compounds.

11.1 Calculation of EAN

An equation for calculating the EAN may be represented as follows:

$$EAN = Z + a + b + c$$

Where,

Z = Atomic number of metal atom

a = Number of electrons donated by terminal carbonyl groups

b = Number of electrons donated by bridging carbonyl groups

c = Number of electrons donated by other metal atoms for the formation of M—M bonds

The EAN for a nickel atom in nickel tetracarbonyl can be calculated as follows:

Z = Atomic number of metal atom = Atomic number of nickel atom = 28

a = Number of electrons donated by terminal carbonyl groups = 4 terminal carbonyl groups × 2 electrons donated by each group = 8

b = Number of electrons donated by bridging carbonyl groups = 0 (because there are no bridge bonds)

c = Number of electrons donated by other metal atoms for the formation of M—M bonds = 0 (Because there are no M—M bonds)

Thus,

$$EAN = Z + a + b + c = 28 + 8 + 0 + 0 = 36$$

Here, the effective atomic number is found to be 36, which is the atomic number of Krypton ($Z = 36$), a noble gas lying in the same period of the periodic table as Ni ($Z = 28$).

Thus, EAN rule is said to be obeyed in nickel tetracarbonyl.

In order to check the validity of the 18-electron rule, we have to put the number of valence electrons of the metal atom in place of the atomic number (Z) in the EAN equation.

The valence shell nickel (Ni = $3d^8\ 4s^2$) atom has 10 valence electrons.

Thus, the number of valence electrons surrounding the nucleus of the metal atom = number of valence electrons of metal atom + $a + b + c$ = $10 + 8 + 0 + 0 = 18$.

Thus, nickel tetracarbonyl obeys the 18-electron rule.

The EAN for iron atoms in triiron dodecacarbonyl can be calculated as follows:

In triiron dodecacarbonyl, two iron atoms are bridged and have same environment, while the third iron atom has a different bonding environment. Thus, we need to calculate the EAN for both the types of iron atoms separately.

For bridged iron atoms:

Z = Atomic number of metal atom = Atomic number of iron atom = 26

a = Number of electrons donated by terminal carbonyl groups = 3 terminal carbonyl groups × 2 electrons donated by each group = 6

b = Number of electrons donated by bridging carbonyl groups = 2 bridging carbonyl groups × 1 electron donated by each group = 2

c = Number of electrons donated by other metal atoms for the formation of M—M bonds = 2 Fe—Fe bonds × 1 electron donated by each Fe atom = 2

Thus,

$$\text{EAN} = Z + a + b + c = 26 + 6 + 2 + 2 = 36$$

Here, the effective atomic number is found to be 36, which is the atomic number of Krypton ($Z = 36$), a noble gas lying in the same period of the periodic table as Fe ($Z = 26$).

Thus, EAN rule is said to be obeyed by the bridging Fe atoms in triiron dodecacarbonyl.

In order to check the validity of the 18-electron rule, we have to put the number of valence electrons of the metal atom in place of the atomic number (Z) in the EAN equation.

The valence shell iron (Fe = $3d^6\ 4s^2$) atom has eight valence electrons.

Thus, the number of valence electrons surrounding the nucleus of the metal atom = number of valence electrons of metal atom + $a + b + c$ = $8 + 6 + 2 + 2 = 18$.

Thus, the bridging Fe atoms in triiron dodecacarbonyl obey the 18-electron rule.

For unbridged iron atom:

Z = Atomic number of iron atom = 26

a = 4 terminal carbonyl groups × 2 electrons donated by each group = 8

b = 0 (Because no bridges are formed)

c = 2 Fe—Fe bonds × 1 electron donated by each Fe atom = 2

Thus,

$$\text{EAN} = Z + a + b + c = 26 + 8 + 0 + 2 = 36$$

Here, the effective atomic number is found to be 36, which is the atomic number of krypton ($Z = 36$), a noble gas lying in the same period of the periodic table as Fe ($Z = 26$).

Thus, EAN rule is said to be obeyed by the non-bridging Fe atom in triiron dodecacarbonyl.

In order to check the validity of the 18-electron rule, we have to put the number of valence electrons of the metal atom in place of the atomic number (Z) in the EAN equation.

The valence shell of a nickel (Fe $= 3d^6\ 4s^2$) atom has 8 valence electrons.

Thus, the number of valence electrons surrounding the nucleus of the metal atom = number of valence electrons of metal atom $+ a + b + c =$ $8 + 6 + 2 + 2 = 18$.

Thus, the non-bridging Fe atom in triiron dodecacarbonyl obeys the 18-electron rule.

The calculations of EAN for some metal carbonyls are summarized in Table 3.

While finding the EAN, if only valence electrons of the metal are considered, the resultant number for stable complexes comes out to be 18. Hence, the EAN rule is now referred to as Langmuir's 18-electron rule.

The octahedral complexes obeying the 18-electron rule (18-electron compounds) are especially stable. In order to understand this, consider the energy level diagram of an octahedral complex in presence of a strong field ligand, shown in Figure 27.

Carbon monoxide is considered as a strong field ligand, because despite its poor ability to donate σ-electrons, it has a remarkable ability to act as a π-acceptor. The symmetry adapted combinations of six σ (a_{1g}, t_{1u} and e_g) and three π (t_{2g}) orbitals of the ligand are shown on the right-hand side of the figure. The t_{2g} set of orbitals of the metal atom also act as bonding orbitals attributed to the presence of π-interactions between the metal and ligand orbitals. In the MO diagram, there are nine bonding molecular orbitals. Thus, compounds containing all these bonding molecular orbitals (BMOs) filled with 18 electrons are very stable.

If more than 18 electrons are to be accommodated, the anti bonding molecular orbitals (ABMOs) must be filled. Such compounds are less stable and readily lose an electron, showing the behaviour of reducing agents. Similarly, compounds with less than 18 electrons will have relatively low stability and the tendency to react further in order to achieve 18-electron configurations.

Table 3 Calculation of effective atomic number (EAN) in some carbonyls

Metal carbonyl	Z	a	b	c	EAN = Z + a + b + c
Ni(CO)$_4$	28	4 × 2 = 8	0	0	36 [Kr]
Fe(CO)$_5$	26	5 × 2 = 10	0	0	36 [Kr]
Ru(CO)$_5$	44	5 × 2 = 10	0	0	54 [Xe]
Os(CO)$_5$	76	5 × 2 = 10	0	0	86 [Rn]
Cr(CO)$_6$	24	6 × 2 = 12	0	0	36 [Kr]
Mo(CO)$_6$	42	6 × 2 = 12	0	0	54 [Xe]
W(CO)$_6$	74	6 × 2 = 12	0	0	86 [Rn]
Fe$_2$(CO)$_9$	26	3 × 2 = 6	3 × 1 = 3	1 × 1 = 1	36 [Kr]
Co$_2$(CO)$_8$ (bridged)	27	3 × 2 = 6	2 × 1 = 2	1 × 1 = 1	36 [Kr]
Co$_2$(CO)$_8$ (without bridge)	27	4 × 2 = 8	0	1 × 1 = 1	36 [Kr]
Mn$_2$(CO)$_{10}$	25	5 × 2 = 10	0	1 × 1 = 1	36 [Kr]
Fe$_3$(CO)$_{12}$ (for un bridged Fe)	26	4 × 2 = 8	0	2 × 1 = 2	36 [Kr]
Fe$_3$(CO)$_{12}$ (for bridged Fe)	26	3 × 2 = 6	2 × 1 = 2	2 × 1 = 2	36 [Kr]

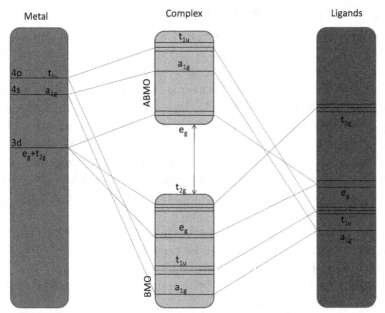

Figure 27 Energy level diagram of an octahedral complex in the presence of a strong field ligand.

Eight BMOs are present in a square planar complex in the presence of a strong field ligand. Thus, a 16-electron configuration is required for a stable square planar complex in the presence of a strong field ligand. However, considering the donation of two electrons from each of the four ligands, only eight electrons can be managed from the ligand side. Due to this, the metal atom must provide an additional eight electrons. This is possible only with the metal ions of group 9 (Co, Rh, Ir) and 10 (Ni, Pd, Pt) lying toward the right-hand side of the d-block.

Most of the organometallic compounds, including carbonyls and nitrosyls, obey the EAN rule.

It is mainly useful in predicting the number of ligands attached to the metal in such compounds.

11.2 Electron Counting Methods

There are two popular methods giving the same results for the electron count. They are the following:

Neutral ligand method (covalent method)

Donor pair method (ionic method)

11.3 Neutral Ligand Method

In this method, all the ligands are treated as electrically neutral. It takes into account the number of electrons it can donate in its neutral state. The neutral ligands capable of donating two electrons are designated as L. The ligands such as Cl^-, which can donate one electron in their neutral state, are designated as X type ligands. The ligand cyclopentadienyl (η^5-C_5H_5), which is a five-electron donor, is designated by a combined symbol L_2X. This method is easy to use when the ligands are properly designated. The overemphasis on the degree of covalence along with negligence of the charge over the metal ion remain shortcomings of this method. Due to this, it becomes difficult to assign oxidation states to the metal ion, resulting in the loss of important information related to the ligands.

The verification of the 18-electron rule for a mixed ligand carbonyl complex (η^5-C_5H_5)Fe(CO)$_2$Cl can be carried out as follows:

In this complex, the Fe atom has eight valence electrons.

In addition to this, the ligand η^5-C_5H_5, when considered as a neutral ligand, contributes five electrons.

CO is a two-electron donor, thus two CO ligands contribute four electrons.

Cl, counted as a neutral species, is a single electron donor, which contributes one electron in total.

Thus, the total electron count can be shown as:

One Fe atom	8 electrons
One (η^5-C_5H_5) ligand (L_2X)	5 electrons
Two CO ligands (L)	4 electrons
One chlorine ligand (X)	1 electron
Total electron count	**18 electrons**

An organometallic compound, containing ligands designated by L and X, can be shown as $[MX_aL_b]^c$, where a is the number of ligands of type X, b is the number of ligands of type L and c is the charge over the complex.

Electron Count $= n + a + 2b - c$, where n is the group number of the metal in the periodic table.

(η^5-C_5H_5)Fe(CO)$_2$Cl can be represented as $[(L_2X)M(2L)(X)]$ or $[MX_2L_4]$.

Electron Count $= n + a + 2b - c = 8 + 2 + 2 \times 4 - 0 = 18$

Table 4 Some ligands and their electron contribution according to 'neutral ligand' and 'donor pair' methods

Ligand	Neutral ligand method	Donor pair method
H	1(X)	2(H$^-$)
F, Cl, Br, I	1(X)	2(X$^-$)
CO	2(L)	2
(η^5-C$_5$H$_5$)	5(L$_2$X)	6(C$_5$H$_5$$^-$)
NO (linear)	3	2(NO$^+$)
NO (Bent)	1	2(NO$^-$)

11.4 Donor Pair Method

According to this method, some ligands are treated as neutral, whereas others are treated as charged. It is assumed that the ligands donate electrons only as pairs. Neutral ligands like CO are considered as two electron donors. Ligands like halides are considered to take an electron from metal and are treated as X$^-$. The ligand (η^5-C$_5$H$_5$) is considered as C$_5$H$_5$$^-$, which becomes a six-electron donor.

The oxidation state of the metal is calculated as total charge over the complex minus charges over the ligands. The number of electrons contributed by metal is calculated as the group number minus its oxidation number. Finally, the electron count is completed as the total of electrons on the metal and the electrons contributed by the ligands.

Calculation for (η^5-C$_5$H$_5$)Fe(CO)$_2$Cl can be done as follows:

Here, oxidation state of Fe, can be calculated as

$$-1 + X + 0 - 1 = 0$$
$$X = +2$$

The group number of Fe is 8.

Therefore, the number of electrons contributed by Fe is $8 - 2 = 6$.

The number of electrons contributed by one C$_5$H$_5$$^-$ = 6.

The number of electrons contributed by two CO = 4.

Number of electrons contributed by one Cl$^-$ = 2.

The number of electrons contributed by some ligands by the neutral ligand method and donor pair method are summarized in Table 4.

12. CATALYTIC ACTIVITY OF METALLIC CARBONYLS

12.1 Common Steps Involved in Homogeneous Catalytic Cycles

Organometallic compounds in general and metal carbonyls in particular are known to show catalytic activity [6,7]. Most of the homogeneous catalytic cycles involve five steps of reaction. Hence, it is imperative to understand these steps before studying the catalytic cycles involved in different syntheses. These steps are listed below:

1. Coordination of a ligand and its dissociation
2. Migratory insertions and β-eliminations
3. Nucleophilic attack on coordinated ligands
4. Oxidations and reductions
5. Oxidative additions and reductive eliminations

12.1.1 Coordination of Ligand and its Dissociation

An efficient catalytic cycle requires a facile entry and exit of the ligand. Both coordination and dissociation of ligands must occur with low-activation free energy. Labile metal complexes are therefore essential in catalytic cycles. Coordinatively unsaturated complexes containing an open or weakly coordinated site are labile.

Square-planner 16-electron complexes are coordinatively unsaturated and are usually employed to catalyse the reactions of organic molecules. Catalytic systems involving ML_4 complexes of Pd (II), Pt (II) and Rh (I), like hydrogenation catalyst [RhCl(PPh$_3$)$_3$], are well known.

12.1.2 Migratory Insertions and β-Eliminations

The migration of an alkyl ligand to an unsaturated ligand is shown in the reaction. This reaction is an example of a migratory insertion reaction.

The migration of a hydride ligand to a coordinated alkene to produce a coordinated alkyl ligand is shown in the reaction as follows:

Elimination is the reverse of insertion. Elimination of β-hydrogen is illustrated in a reaction shown as follows:

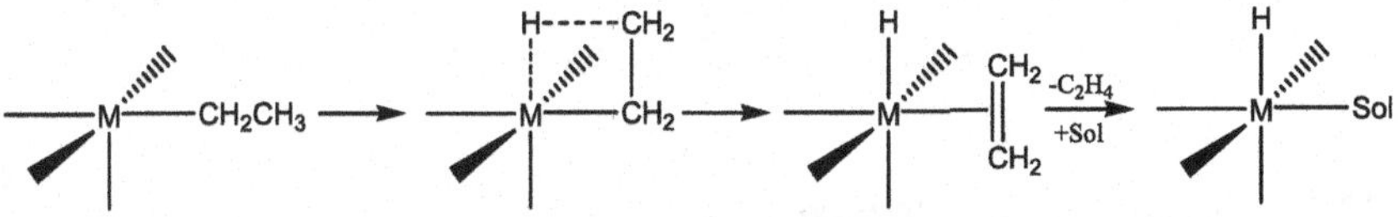

12.1.3 Nucleophilic Attack on Coordinated Ligands

The coordination of ligands, such as carbon monoxide, and alkenes to metal ions in positive oxidation states activates the coordinated C atoms for a nucleophilic attack. These reactions have found special attention in catalysis, as well as organometallic chemistry.

The hydration of coordinated ethylene with Pd (II) is an example of catalysis by nucleophilic activation. Stereochemical evidences indicate that the reaction occurs by direct attack on the most highly substituted carbon atom of the coordinated alkene:

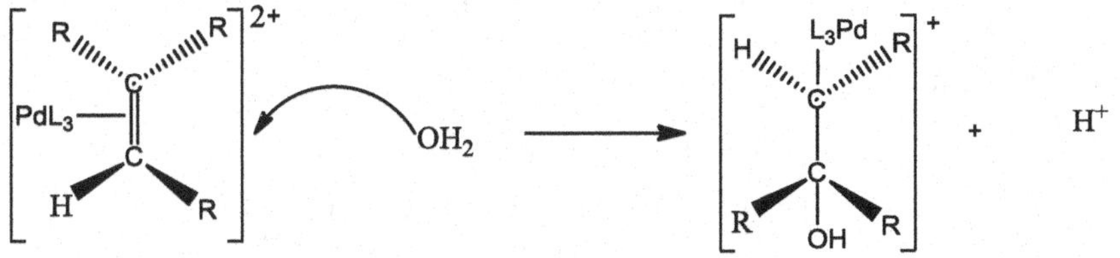

The hydroxylation of a coordinated alkene can also occur by the coordination of H_2O ligand to a metal complex followed by an insertion reaction as follows:

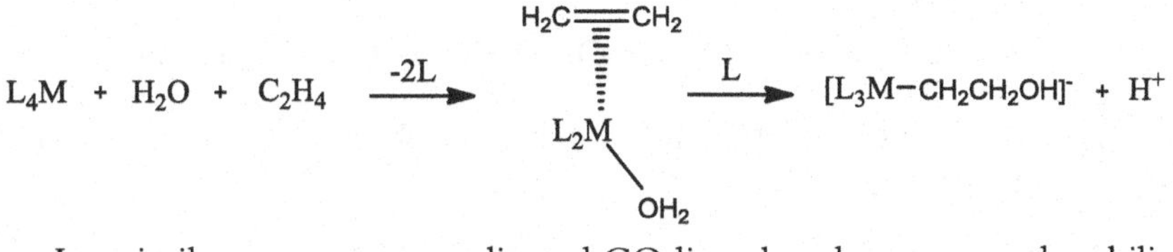

In a similar manner, a coordinated CO ligand undergoes a nucleophilic attack by an OH^- ion at the C atom, forming a $-CO(OH)$ ligand, which subsequently loses CO_2 as follows:

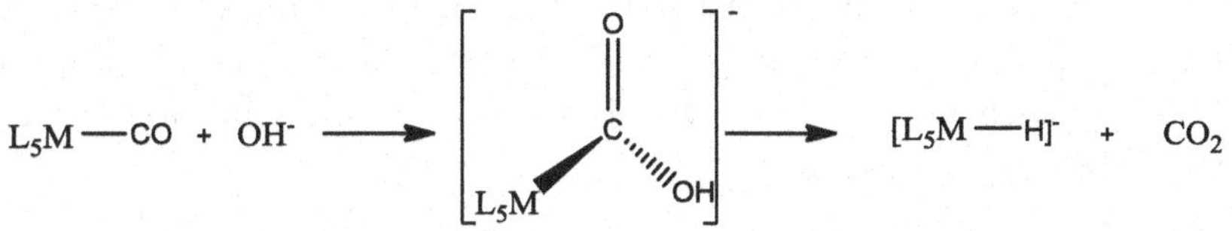

The water-gas shift reaction catalysed by metal carbonyl complexes or metal ions on solid surfaces involves this step:

$$CO + H_2O \xrightarrow{\text{catalyst}} CO_2 + H_2$$

12.1.4 Oxidation and Reduction

The use of metal complexes in the catalytic oxidation of organic compounds is popular. In these catalytic cycles, the metal atom shuttles between two oxidation states. Examples of common catalytic one-electron couples are Cu^{2+}/Cu^{+}, Co^{3+}/Co^{2+} and Mn^{3+}/Mn^{2+}. An example of a catalytic two-electron couple is Pd^{2+}/Pd.

Catalysts containing metal ions are also used in oxidation of hydrocarbons. The oxidation of *p*-xylene to terephthalic acid is an example of such use. The metal ions play different roles in these radical oxidations, as

Initiation (In˙ = Initiator radical):

$$In^\bullet + R{-}H \rightarrow In{-}H + R^\bullet$$

Propagation:

$$R^\bullet + O_2 \rightarrow R{-}O{-}O^\bullet$$
$$R{-}O{-}O^\bullet + R{-}H \rightarrow R{-}O{-}O{-}H + R^\bullet$$

Termination:

$$2R^\bullet \rightarrow R_2$$
$$R^\bullet + R{-}O{-}O^\bullet \rightarrow R{-}O{-}O{-}R$$
$$2R{-}O{-}O^\bullet \rightarrow R{-}O{-}O{-}O{-}O{-}R \rightarrow O_2 + \text{nonradicals}$$

The metal ions control the reaction by contributing to the formation of the $R{-}O{-}O{-}$ radicals:

$$Co(II) + R{-}O{-}O{-}H \rightarrow Co(ROOH) \rightarrow Co(III)OH + R{-}O^\bullet$$
$$Co(III) + R{-}O{-}O{-}H \rightarrow Co(II) + R{-}O{-}O^\bullet + H^+$$

The metal atom shuttles back and forth between the two oxidation states in this pair of reactions. A metal ion can also act as an initiator. For example, when arenes are involved, it is thought that the initiation occurs by a simple redox process as

$$ArCH_3 + Co(III) \rightarrow ArCH_3^\bullet + Co(II)$$

12.1.5 Oxidative Addition and Reductive Elimination

The oxidative addition of a molecule AX to a complex occurs by the dissociation of the A—X bond followed by the coordination of two fragments as

The reductive elimination is the reverse of the oxidative addition and generally follows it in a catalytic cycle.

The mechanisms of oxidative addition reactions are different. Depending upon reaction conditions and the nature of the reactants, there is evidence for oxidative addition by a simple concerted reaction, heterolytic (ionic) addition of A^+ and X^-, or radical addition of $A^{\bullet}$ and $X^{\bullet}$. Even though it has been observed that the rates of oxidative addition of alkyl halides generally follow the orders:

Primary alkyl < secondary alkyl < tertiary alkyl

F << Cl < Br < I

12.2 Catalytic Hydrocarbonylation

Dicobalt octacarbonyl is employed as a catalyst precursor in the hydrocarbonylation of alkenes to produce aldehydes.

At 150 °C and 200 atmospheric pressure, the dicobalt octacarbonyl reacts with hydrogen to establish equilibrium with a tetracarbonyl hydrido complex.

$$Co_2(CO)_8 + H_2 \rightarrow 2[CoH(CO)_4]$$

The hydrido complex loses a carbon monoxide ligand to initiate the catalytic cycle of hydrocarbonylation, shown in Figure 28. The $CoH(CO)_3$ thus produced coordinates an alkene to give [CoH(CO)$_3$(CH$_2$= CHCH$_2$CH$_3$)], which undergoes an insertion reaction with the coordinated hydrido ligand and again coordinates the carbon monoxide ligand. The alkyl complex thus produced undergoes a migratory insertion reaction and coordinates yet another carbon monoxide ligand to yield an acyl complex, as shown in Figure 28. The hydrocarbonylation occurs due to the attack by H_2 to give the product. The cycle continues due to the regeneration of the catalyst at the end of the last step.

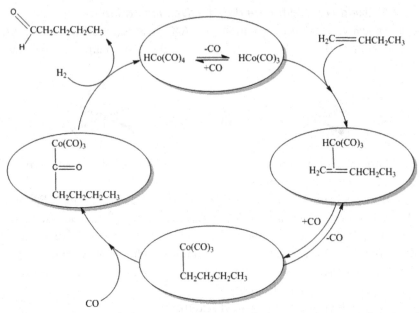

Figure 28 Catalytic cycle for hydrocarbonylation of 1-butene using a cobalt carbonyl catalyst.

12.3 The Pauson–Khand Reaction

The dicobalt octacarbonyl catalysed the three-component reaction of alkene, alkyne and carbon monoxide to give cyclopentenone, which was an important breakthrough in organic synthesis [8]. A three-component direct condensation to yield rings or fragments was very difficult earlier to the establishment of the Pauson–Khand reaction.

Stirring a solution of bicyclo[2.2.1]hept-2-ene and dicobalt octacarbonyl in isooctane with acetylene gas at 60–70 °C followed by passing a 1:1 gaseous mixture of acetylene and carbon monoxide yields the product (Endo 3a,4,5,6,7,7a-Hexahydro-4,7-methano-2-indene-1-one) shown in the reaction.

The general mechanism of these reactions involves the formation of cobalt complexes with acetylene followed by the insertion of alkene and carbonyl to give an acyl cobalt complex. This complex then undergoes a reductive elimination to produce cyclopenetone.

12.4 Catalytic Carbonylation of Methanol

12.4.1 Monsanto Acetic Acid Synthesis

Rhodium-catalysed carbonylation of methanol has become a very successful commercial process [9]. This process for the manufacture of acetic acid is popularly known as the Monsanto process.

$$CH_3OH + CO \xrightarrow{\left[RhI_2(CO)_2\right]} CH_3COOH$$

All three members of group nine (cobalt, rhodium, and iridium) catalyse this reaction. The complexes of the 4d metal, rhodium, are found to be the most active. Earlier, a cobalt complex was used for this reaction, but the rhodium catalyst developed at Monsanto greatly reduced the cost of the process by allowing lower pressures to be used.

The catalytic cycle in the Monsanto process is shown in Figure 29.

The cycle begins with the slowest step, involving the oxidative addition of iodomethane to the four-coordinate, 16-electron complex $[RhI_2(CO)_2]^-$ to give a six-coordinate 18-electron complex $[(H_3C)RhI_3(CO)_2]^-$, as shown in the catalytic cycle. The migratory insertion of carbon monoxide then yields a 16-electron acetyl complex $[RhI_3(CO)(COCH_3)]^-$. The coordination of carbon monoxide restores an 18-electron complex, which then undergoes reductive elimination of acetyl iodide, generating $[RhI_2 (CO)_2]^-$. Finally, water hydrolyses the acetyl iodide to acetic acid and regenerates HI as the following:

$$CH_3COI + H_2O \rightarrow CH_3COOH + HI$$

Iodide is the most suitable anion in this catalytic system, as it undergoes the oxidative addition of iodomethane much faster than any other haloalkanes. In addition to this, the soft I^- ion is a good ligand for the soft Rh (I), which probably forms a five-coordinate complex, $[RhI_3(CO)_2]^{2-}$, which undergoes oxidative addition with iodomethane much faster than $[RhI_2 (CO)_2]^-$

The strong acid HI is also effective in halogenating methanol:

$$CH_3OH + HI \rightarrow CH_3I + H_2O$$

The Monsanto process has at least two drawbacks. The iodide moiety used in this process is corrosive. The rhodium complex used in the process

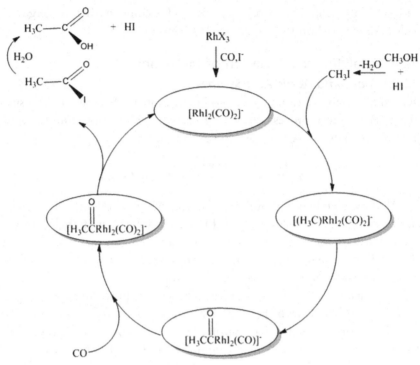

Figure 29 Catalytic cycle for carbonylation of methanol by Monsanto acetic acid synthesis.

is stable only under high pressures of carbon monoxide, which makes the recycling of the catalyst difficult.

13. EXERCISES

13.1 Multiple Choice Questions

1. Majority of metal carbonyls are _____.
   (a) paramagnetic   (b) diamagnetic
   (c) ferromagnetic   (d) none of these
2. Neutral Carbonyl hydrides are _____.
   (a) acidic   (b) basic
   (c) neutral   (d) amphoteric
3. Anionic Carbonyl hydrides are _____.
   (a) acidic   (b) basic
   (c) neutral   (d) amphoteric

4. Which of the following is used in Mond's process?
(a) $Ni(CO)_4$ (b) $Co_2(CO)_8$
(c) $Cr(CO)_6$ (d) $Fe(CO)_5$

5. The hybridization on Fe atom of $Fe(CO)_5$ is?
(a) dsp^3 (b) d^2sp^3
(c) sp^3d^2 (d) sp^3d

6. How many nitrosyls are required to replace three carbonyl groups of a metal carbonyl?
(a) 2 (b) 3
(c) 1 (d) 4

7. Which catalyst is used in the Pauson–Khand reaction?
(a) $Co_2(CO)_8$ (b) $Fe_3(CO)_{12}$
(c) $Fe_2(CO)_9$ (d) $Mn_2(CO)_{10}$

8. Which one of the following catalyses Monsanto acetic acid synthesis?
(a) $[Rh_4(CO)_{16}]$ (b) $[RhI_2(CO)_2]^-$
(c) $Co_2(CO)_8$ (d) $Fe_3(CO)_{12}$

13.2 Short/Long Answer Questions

1. 'Vanadium hexacarbonyl is paramagnetic': Explain
2. Describe the Heiber base reaction.
3. How is the C—O bond order related to the C—O stretching frequencies in metal carbonyls? Explain.
4. Calculate the EAN for $V(CO)_6$.
5. Explain the acidic and basic behaviour of carbonyl hydrides.
6. How is the Collman's reagent prepared?
7. Discuss the preparation, properties and structure of dicobalt octacarbonyl.

SUGGESTED FURTHER READINGS

The topics discussed in this chapter are a part of a standard graduate curriculum. A majority of the textbooks with titles related to coordination chemistry and organometallic chemistry can act as a source of further reading. Moreover, there are several web resources useful for further learning. Some of them are listed below.

https://en.wikipedia.org/wiki/Metal_carbonyl
http://nptel.ac.in/courses/104106064/lectures.pdf
http://www.chem.tamu.edu/rgroup/marcetta/chem462/lectures/Lecture%204%20%20Metal%20Carbonyls.pdf
www.technology.matthey.com/pdf/pmr-v16-i2-050-055.pdf
http://www.britannica.com/science/metal-carbonyl
http://www.york.ac.uk/media/chemistry/research/douthwaite/webM-L%20and%20M-M%20bonding%202011-12.pdf
http://textofvideo.nptel.iitm.ac.in/104108062/lec3.pdf

REFERENCES

The following books and research articles contain advanced topics and nuances in the chemistry of metal carbonyls:

[1] Behrens H. The chemistry of metal carbonyls: "the life work of Walter Hieber". J Organomet Chem 1975;94(2):139–59.
[2] House JE. Chapter 21-complexes containing Metal–Carbon and Metal–Metal bonds. In: House JE, editor. Inorganic chemistry. 2nd ed. Academic Press; 2013. p. 707–46.
[3] Schatzschneider U. PhotoCORMs: light-triggered release of carbon monoxide from the coordination sphere of transition metal complexes for biological applications. Inorganica Chim Acta 2011;374(1):19–23.
[4] House JE. Chapter 22-coordination compounds in catalysis. In: House JE, editor. Inorganic chemistry. 2nd ed. Academic Press; 2013. p. 747–72.
[5] Ponec R. Structure and bonding in binuclear metal carbonyls. Classical paradigms vs insights from modern theoretical calculations. Comput Theor Chem 2015;1053: 195–213.
[6] Ramachandran R, et al. Ruthenium(II) carbonyl complexes designed with arsine and PNO/PNS ligands as catalysts for N-alkylation of amines via hydrogen autotransfer process. J Organomet Chem 2015;791:130–40.
[7] Mkoyi HD, et al. (Pyrazol-1-yl)carbonyl palladium complexes as catalysts for ethylene polymerization reaction. J Organomet Chem 2013;724:95–101.
[8] Park JH, Chang K-M, Chung YK. Catalytic Pauson–Khand-type reactions and related carbonylative cycloaddition reactions. Coord Chem Rev 2009;253(21–22):2461–80.
[9] Haynes A. 6.01-Carbonylation reactions. In: Poeppelmeier JR, editor. Comprehensive inorganic chemistry II. 2nd ed. Amsterdam: Elsevier; 2013. p. 1–24.

CHAPTER 9

Metal Nitrosyls

Contents

Essentials of Coordination Chemistry
http://dx.doi.org/10.1016/B978-0-12-803895-6.00009-4

1. INTRODUCTION

Metal nitrosyls are the transition metal complexes of nitric oxide (NO) containing a metal–nitrogen bond. Roussin's red salt, $Na_2[Fe_2(NO)_4S_2]$, and Roussin's black salt, $Na[Fe_4(NO)_7S_3]$, were the earliest known metallic nitrosyls. In line with the inclusion of metal carbonyl complexes under the category of organometallic compounds, the metallic nitrosyl complexes are also recently included in organometallic compounds. The nitric oxide cation is isoelectronic with carbon monoxide. Hence, there is quite a bit of resemblance in the chemistry of metallic carbonyls and nitrosyls. However, the contrasts in these chemistries are also noteworthy.

Out of the five valence electrons, the nitrogen of nitric oxide utilizes two of them for the formation of a double bond with oxygen. The remaining three electrons remain on nitrogen as a lone pair and an odd electron. This makes nitric oxide a ligand of interest from the vantage point of structure, bonding and reactivity of its complexes [1].

2. BONDING IN METALLIC NITROSYLS

2.1 Nitric Oxide

In order to understand the bonding in metallic nitrosyls, let us first see the molecular orbital (MO) diagram of a nitric oxide (NO) molecule, shown in Figure 1.

The order of energy of the molecular orbitals and the accommodation of 11 electrons of the nitric oxide can be shown as follows:

$$\left(\sigma_s^{\,b}\right)^2 \left(\sigma_p^{\,b}\right)^2 \left(\pi_y^{\,b} = \pi_z^{\,b}\right)^4 \left(\sigma_s^{\,*}\right)^2 \left(\pi_y^{\,*} = \pi_z^{\,*}\right)^1 \left(\sigma_p^{\,*}\right)^0$$

The following steps are involved during the formation of a linear metallic nitrosyl:

The odd electron on nitrogen of nitric oxide $\left(\pi_y^{\,*} = \pi_z^{\,*}\right)^1$ is lost to form nitrosyl cation, wherein a lone pair of electrons from oxygen is transferred to form the third bond between nitrogen and oxygen.

$$:\ddot{O}\!=\!\!=\!\dot{\underset{..}{N}} \quad \longrightarrow \quad \left[:O\!=\!\!\equiv\!\!N:\right]^+ + \bar{e}$$

The odd electron $\left(\pi_y^{\,*} = \pi_z^{\,*}\right)^1$ is gained by the metal atom (M^0), which decreases its formal oxidation state (M^-) by one unit.

$$M^0 + \bar{e} \quad \longrightarrow \quad M^-$$

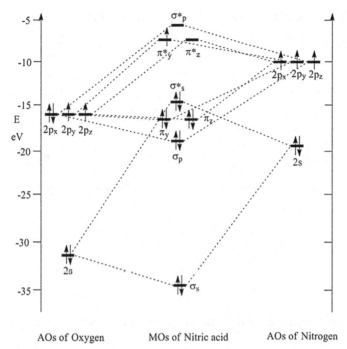

AOs of Oxygen   MOs of Nitric acid   AOs of Nitrogen

Figure 1 Molecular orbital energy level diagram of nitric oxide.

Nitrogen of the NO^+ (nitrosyl or nitrosonium ion) donates an electron pair to the metal ion (M^-) for coordination.

$$\left[:O\!\!\equiv\!\!N:\right]^+ + M^- \longrightarrow \left[:O\!\!\equiv\!\!N\!\longrightarrow\!M^-\right]$$

A majority of the nitrosyl complexes contain NO^+ (nitrosyl or nitrosonium ion) as the ligand. The nitrosyl cation is isoelectronic with carbon monoxide. Hence, the bonding between nitrogen and metal in nitrosyls is analogous to that of carbon and metal in carbonyls. However, the ligand nitrosyl (NO^+) is a three-electron donor, as it donates one electron to the metal ion prior to the donation of an electron pair for the formation of a coordinate covalent bond.

2.2 Formation of Dative σ-Bond

The overlapping of an empty hybrid orbital (a blend of d-, s- and p-orbitals) of a metal atom with the filled hybrid orbital (HOMO) on a nitrogen atom of the (NO^+) ion results into the formation of an M ← NO^+ σ-bond, as shown in Figure 2.

2.3 Formation of π-Bond by Back-Donation

This bond is formed as a result of the overlapping of filled $d\pi$ orbitals or hybrid $dp\pi$ orbitals of a metal atom with low-lying empty (LUMO) orbitals on NO^+ ion, i.e. $M \xrightarrow{\pi} NO^+$, as shown in Figure 3.

From the MO diagram of a nitric oxide molecule, as shown in Figure 4, we can see that σ_s^* is the highest occupied molecular orbital (HOMO) after

Figure 2 Formation of an $M \leftarrow NO^+$ σ-bond in metal nitrosyls.

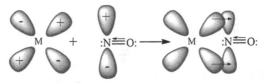

Figure 3 Formation of $M \xrightarrow{\pi} NO^+$ bond by back-donation in metal nitrosyls.

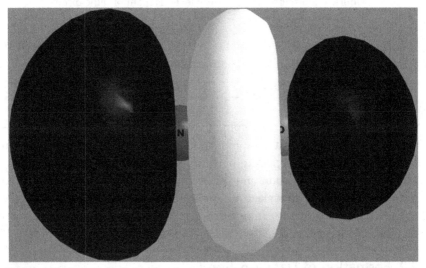

Figure 4 Highest occupied molecular orbital (HOMO) of a nitrosyl ion (Red (light grey in print versions) colour is for the positive sign of the wave function, while the blue (dark grey in print versions) colour indicates the negative sign of the wave function).

Figure 5 Lowest unoccupied molecular orbital (LUMO) of a nitrosyl ion (Red (light grey in print versions) colour is for the positive sign of the wave function, while the blue (dark grey in print versions) colour indicates the negative sign of the wave function).

the loss of one electron. This orbital can donate the lone pair of electrons for the formation of the ON → M σ-bond.

Whereas $(\pi_y^* = \pi_z^*)$ are the lowest unoccupied molecular orbitals (LUMO) that can accept the electron density from an appropriately oriented filled metal orbital, resulting into the formation of an M → NO^+ π-bond, as shown in Figure 5.

In certain nitrosyls such as $[Co^{+3}(CN)_5(NO)]^{-3}$ and $[Co^{+3}(NH_3)_5(NO)]^{+2}$, the nitric oxide is known to exist as a nitric oxide anion (NO^-).

The following steps are involved during the formation of a metallic nitrosyl with a nitric oxide anion (NO^-):

The nitrogen of nitric oxide gains one electron from a metal atom/ion to form a nitric oxide anion (NO^-). Nitrogen of the nitric oxide anion (NO^-) donates an electron pair to the metal for coordination.

$$:\!\overset{\cdot}{N}\!\!=\!\!O \xrightarrow{+e^-} [:\!\overset{\cdot}{N}\!\!=\!\!O]^- \xrightarrow{+M} M\!-\!\overset{\cdot\cdot}{N}$$
$$\underset{O}{\overset{\|}{}}$$

The nitrogen atom of the ligand in this case retains one lone pair. Because of this reason, the metallic nitrosyls containing a nitric oxide anion (NO^-) are bent with an M—N—O bond angle in the range of 120–140 °C. In these types of complexes, the nitric oxide effectively donates only one electron to the metal ion. Hence, NO^- can be considered as one electron donor.

3. INFRARED SPECTROSCOPY

The free nitric oxide has a stretching frequency of 1870 cm^{-1}. A change in this frequency is expected upon ligation. Some of the factors that affect this change are the nature of co-ligands, the charge on the complex and the structure of the complex. We have already seen that the nitrosyl ligand is linear ($\angle$M—N—O = 180°) when it coordinates as NO^+. The N—O stretching frequency in such complexes is observed in a broad range of 1950–1450 cm^{-1}. The nitrosyl ligand is bent when it coordinates as NO^-. The angle M—N—O in such cases has been observed in the range of 120° to 140°. The observed N—O stretching frequency in such complexes is found to be slightly lower, in the range of 1720–1400 cm^{-1}.

A five-coordinated complex [CoCl$_2$(NO)(PCH$_3$Ph$_2$)$_2$], shown in Figure 6, is known to exist in two isomeric forms. The trigonal bipyramidal isomer exhibits N—O stretching frequency at 1750 cm^{-1}, while the square pyramidal isomer absorbs at 1650 cm^{-1}. The bond angle and nitrosyl-stretching frequencies of some mononuclear nitrosyl complexes of the first transition series is shown in Table 1.

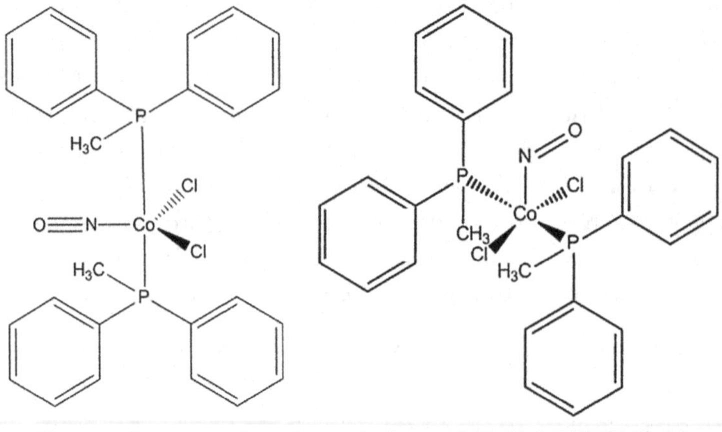

Figure 6 Structure of two isomeric forms of [CoCl$_2$(NO)(PCH$_3$Ph$_2$)$_2$].

Table 1 Comparison N—O stretching in some mononuclear metallic nitrosyls

Nitrosyl	Type	∠M—N—O	ν N—O (cm^{-1})
Nitric oxide	Free	Not applicable	1870
[Fe(CN)$_5$NO]$^{-2}$	Cyano nitrosyl	~175	1940
[Mn(CN)$_5$NO]$^{-3}$	Cyano nitrosyl	~175	1725
[Cr(CN)$_5$NO]$^{-4}$	Cyano nitrosyl	~175	1515
[V(CN)$_5$NO]$^{-1}$	Cyano nitrosyl	~175	1530
[Coen$_2$Cl(NO)]$^{+1}$	Nitric oxide as NO$^-$	121	1611
[Co(NH$_3$)$_5$(NO)]$^{+2}$	Nitric oxide as NO$^-$	119	1610

There is a significant overlapping in the above shown ranges. Due to this, a correlation between bond angle and N—O stretching frequency is not possible in these complexes. The complexes containing a bridging nitrosyl ligand exhibit the N—O stretching frequency around 1650–1300 cm^{-1}. The bridging nitrosyl ligands are also regarded as linear nitrosyls. Complexes with doubly and triply bridging nitrosyls are known. Triply bridging nitrosyls absorb at frequencies lower than that of doubly bridging ones.

A bridged binuclear complex of chromium, [(h^5-C$_5$H$_5$)(NO)Cr(μ-NO) (μ-NH$_2$)Cr(NO)(h^5-C$_5$H$_5$)], is shown in Figure 7. The structure contains two terminal M—N—O groups and one double bridging μ-NO group.

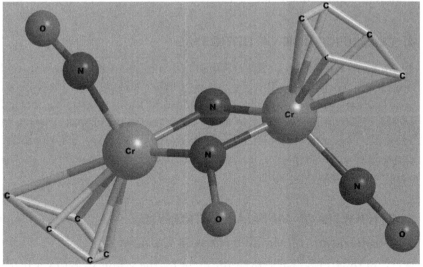

Figure 7 Structure of (h^5-C$_5$H$_5$)(NO)Cr(μ-NO)(μ-NH$_2$)Cr(NO)(h^5-C$_5$H$_5$).

Table 2 Different types of metal nitrosyls

Type	Examples
Nitrosyl carbonyls	$[Co^-(NO^+)(CO)_3]^0$, $[Fe^{2-}(NO^+)_2(CO)_2]^0$, $[Mn^{3-}(NO^+)_3(CO)]^0$, $[Mn^-(NO^+)(CO)_4]^0$, $[V^-(NO^+)(CO)_5]^0$
Nitrosyl halides	$[Fe^-(NO^+)_2I]_2$, $[Fe^-(NO^+)_2I]^0$, $[Fe^0(NO^+)I]^0$, $[Fe^{-2}(NO^+)_3Cl]^0$, $[Co^-(NO^+)_2X]^0$ (X = Cl, Br, I)
Nitrosyl thio complexes	$M^+[Fe^-(NO^+)_2S]^-$, $M^+[Co^-(NO^+)_2S]^-$, $M^+[Ni^-(NO^+)_2S]^-$
Nitrosyl cyano complexes	$[Mn^+(NO^+)(CN)_5]^{2-}$, $[Fe^+(NO^+)(CN)_5]^{2-}$, $[Mn^+(NO^+)(CN)_5]^{3-}$, $[Mo^+(NO^+)(CN)_5]^{4-}$
Other nitrosyl complexes	$[Co^+(NO^+)(NH_3)_5]^{2+}$, $[Co^+(NO^+)(NO_2)_5]^{3-}$, $[Fe^+(NO^+)]^{2+}$, $[Ru^{2+}(NO^+)(NH_3)_4Cl]^{2+}$, $[Ru^{2+}(NO^+)Cl_5]^{2-}$, $[Fe^{2-}(NO^+)_2(PR_3)_3]^0$

In this complex, the band due to terminal nitrosyls is seen at $1644 \ cm^{-1}$. The bridged nitrosyl groups absorb at a lower frequency of $1505 \ cm^{-1}$.

The complex $[(h^5\text{-}C_5H_5)_3Mn_3(NO)_4]$ contains one triply bridging nitrosyl group and three double bridging nitrosyl groups. Two bands corresponding to the doubly bridging groups are observed at $1543 \ cm^{-1}$ and $1481 \ cm^{-1}$, respectively. As expected, the triply bridging nitrosyl absorbs at a lower frequency of $1320 \ cm^{-1}$.

4. CLASSIFICATION OF NITROSYLS

The nitrosyl complexes can be classified into mononitrosyl and polynitrosyl complexes. Each of these types can be further classified on the basis of coordination numbers. A broad classification of the metallic nitrosyl complexes, covering the types of entities included in the chapter, is given in Table 2. Recently, the chemistry of metallic nitrosyls has expanded significantly, which makes it imperative to add more categories to the classification provided here.

5. SYNTHESIS OF METALLIC NITROSYLS

5.1 Preparation of Metallic Nitrosyl Carbonyls

Metal nitrosyl carbonyls can be obtained by the action of NO on metal carbonyls.

$$Fe(CO)_5 + 2NO \xrightarrow{95°C} Fe(CO)_2(NO)_2 + 3CO$$
$$3Fe(CO)_9 + 4NO \rightarrow 2Fe(CO)_2(NO)_2 + Fe(CO)_5 + Fe_3(CO)_{12} + 6CO$$
$$Fe_3(CO)_{12} + 6NO \xrightarrow{85\ °C} 3Fe(CO)_2(NO)_2 + 6CO$$
$$Co_2(CO)_8 + 2NO \xrightarrow{40\ °C} 2Co(CO)_3(NO) + 2CO$$

5.2 Properties of Metal Nitrosyl Carbonyls
5.2.1 Substitution Reactions
Since the nitrosyl ligand is more firmly attached to the metal ion, as compared to the carbonyl group, only carbonyl groups can easily be replaced by ligands like PR_3, CNR and phen.

$$Fe(CO)_2(NO)_2 + 2L(L = PR_3, CNR) \rightarrow Fe(L)_2(NO)_2 + 2CO$$
$$Fe(CO)_2(NO)_2 + phen \rightarrow Fe(phen)(NO)_2 + 2CO$$

5.2.2 Reaction with Halogens
Halogens can convert carbonyl nitrosyls to metal nitrosyl halides.

$$2\left[Fe(CO)_2(NO)_2\right] + I_2 \rightarrow \left[Fe(NO)_2I\right]_2 + 4CO$$

5.3 Preparation of Metallic Nitrosyl Halides
Metal nitrosyl halides can be prepared by the action of nitric oxide on metal halides using a suitable halogen acceptor.

$$CoX_2 + 4NO + Co \rightarrow 2\left[Co(NO)_2X\right]$$
$$2NiI_2 + 2NO + Zn \rightarrow 2[Ni(NO)I] + ZnI_2$$

They can also be prepared by halogenation of nitrosyl carbonyls.

$$2\left[Fe(CO)_2(NO)_2\right] + I_2 \rightarrow \left[Fe(NO)_2I\right]_2 + 4CO$$

5.4 Properties of Metal Nitrosyl Halides
Metal nitrosyl halides react with other ligands to form mononuclear complexes.

$$\left[Fe(NO)_2X\right]_2 + 2L \rightarrow 2\left[Fe(NO)_2(X)(L)\right]$$

Iron nitrosyl halide reacts with potassium sulphide to form a dark red compound with the composition $K_2[Fe_2(NO)_4S_2]$. This compound can be

Figure 8 Structure of Roussin's red salt.

reacted with methyl chloride to give a compound with the composition $K_2[Fe_2(NO)_4(SCH_3)_2]$. Both these compounds are known as Roussin's red salts, shown in Figure 8. In these compounds, Fe is in -1 oxidation state.

$$[Fe(NO)_2I] \xrightarrow{+K_2S} K_2[Fe_2(NO)_4S_2] \xrightarrow{+CH_3Cl} K_2[Fe_2(NO)_4(SCH_3)_2]$$

Roussin's black salt, shown in Figure 9, is a compound with a molecular formula of $NaFe_4S_3(NO)_7$, which is obtained upon mild acidification of the red salt. The red salt can be brought back by alkalization.

Both the Roussin salts find applications as nitric oxide donor in medicinal and food processing fields.

6. INDIVIDUAL NITROSYLS

6.1 $Na_2[Fe^{2+}(CN)_5(NO^+)]$, Sodium Pentacyanonitrosylferrate(II), also known as Sodium Nitroprusside

6.1.1 Preparation
It is prepared by the action of sodium nitrite on sodium ferrocyanide.

$$Na_4[Fe^{2+}(CN)_6] + NaNO_2 + H_2O$$
$$\rightarrow Na_2[Fe^{2+}(CN)_5(NO^+)] + 2NaOH + NaCN$$

Figure 9 Structure of Roussin's black salt.

It can also be prepared by passing nitric oxide in a solution of sodium ferrocyanide in an acidic medium.

$$2Na_2\left[Fe(CN)_6\right] + H_2SO_4 + 3NO$$

$$\rightarrow 2Na_2\left[Fe(NO)(CN)_5\right] + 2NaCN + Na_2SO_4 + \frac{1}{2}N_2 + H_2O$$

It has also been prepared from potassium ferrocyanide. Upon boiling a solution of potassium ferrocyanide with concentrated nitric acid, a mixture of potassium nitrate, potassium ferricyanide and nitric oxide is produced. Subsequent to the crystallization and removal of potassium nitrate, the neutralization of the reaction mixture yields sodium nitroprusside.

$$3K_4\left[Fe(CN)_6\right] + 4HNO_3 \rightarrow 3K_3\left[Fe(CN)_6\right] + NO + 3KNO_3 + 2H_2O$$

$$2K_3\left[Fe(CN)_6\right] + 2NO + 3Na_2CO_3 \rightarrow 2Na_2\left[Fe(CN)_5(NO)\right] + NaOH$$

Figure 10 Structure of sodium nitroprusside.

6.1.2 Structure

The sodium nitroprusside, shown in Figure 10, has an octahedral ferrous ion centre surrounded by five cyano ligands and one nitrosyl ligand. In $[Fe(CN)_5(NO)]^{2-}$, the total positive charge is three units $[Fe^{+2}$ and $NO^+]$, and the negative charge is five units, created due to five cyano groups. Hence, the resultant charge over the ion is -2.

6.1.3 Properties

Sodium nitroprusside forms beautiful ruby red rhombic crystals, which are soluble in water and practically insoluble in acetone, ether and chloroform. It is diamagnetic, which proves that the nitric oxide is present as a nitrosyl ion in this complex. It reacts with sodium hydroxide to give sodium ferrocyanide as

$$6Na_2\left[Fe(CN)_5(NO)\right] + 14NaOH$$
$$\rightarrow 5Na_4\left[Fe(CN)_6\right] + Fe(OH)_2 + 6NaNO_3 + 6H_2O$$

A freshly prepared solution of sodium nitroprusside gives the following colourations:

Violet colouration in presence of sulphide ions as

$$Na_2S + Na_2\left[Fe^{2+}(CN)_5(NO^+)\right] \rightarrow Na_4\left[Fe^{2+}(CN)_5(NO^+)(S^{2-})\right]$$

Red colouration with sulphites of alkali metals as

$$Na_2SO_3 + Na_2\left[Fe(CN)_5(NO)\right] \rightarrow Na_4\left[Fe(CN)_5(NO)(SO_3)\right]$$

Flesh colouration with silver nitrate as

$$2AgNO_3 + Na_2\left[Fe(CN)_5(NO)\right] \rightarrow Ag_2\left[Fe(CN)_5(NO)\right] + 2NaNO_3$$

Deep red colour with aldehydes and ketones containing the CH_3-CO-R group in excess of sodium hydroxide.

6.1.4 Uses

It is used as a reagent in qualitative analysis for the detection of sulphides, sulphites, aldehydes and ketones containing the CH_3-CO-R group. It is also used as a reference compound for the calibration of a Mossbauer spectrometer. Sodium nitroprusside is used to detect illicit substances, such as secondary amines, in forensic laboratories (Simon's test). It is used to control blood pressure and to treat hypertension.

6.2 $[Fe^+(NO^+)(H_2O)_5]SO_4$, Nitrosoferrous Sulphate

It can be prepared by reacting a metal nitrate and a ferrous sulphate in the presence of a few drops of sulphuric acid. The reaction occurs in following three steps:

Metal nitrate reacts with sulphuric acid to produce nitric acid.

$$6NaNO_3 + H_2SO_4 \rightarrow NaHSO_3 + HNO_3$$

Nitric acid oxidizes the ferrous ion (light green) into a ferric ion (reddish brown) and gets reduced to nitric oxide.

$$6FeSO_4 + 2HNO_3 + 3H_2SO_4 \rightarrow 3Fe_2(SO_4)_3 + 2NO + 4H_2O$$

Nitric oxide is absorbed by unreacted ferrous sulphate to form a dark brown ring nitroso ferrous sulphate.

$$FeSO_4 + NO \rightarrow \left[Fe^+(NO^+)(H_2O)_5\right]^{2+}SO_4^{2-}$$

In an aqueous solution, nitroso ferrous sulphate is in penta aqua form $[Fe(NO)(H_2O)_5]^{2+}$. It is a paramagnetic complex with three unpaired electrons showing a magnetic moment of 3.90 bohr magneton (BM). This value of magnetic moment supports the +1 oxidation state of iron (d^7) in a high-spin complex ion.

The formation of a brown ring is a well-known observation for the detection of nitrate in inorganic salts.

6.3 [Cu(CH₃NO₂)₅(NO)][PF₆]₂

A.M. Wright and co-workers have recently synthesized and structurally characterized a copper nitrosyl complex, $[Cu(CH_3NO_2)_5(NO)][PF_6]_2$, which is the first of its kind [2]. The addition of $NOPF_6$ to powdered metallic copper in nitromethane results in the formation of the said complex, shown in Figure 11. The corners of an octahedron are occupied by five CH_3NO_2 ligands and one NO^- ligand in this complex ion. The observed Cu—N—O bond angle in the complex is 121°, which suggests that the nitric oxide is bound as NO^- ion in this complex. The Cu—N

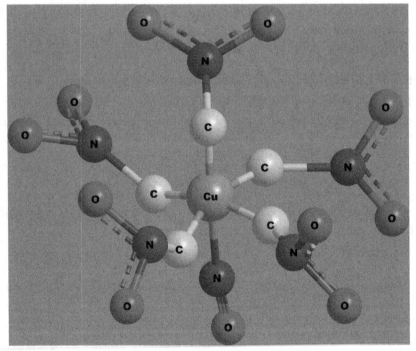

Figure 11 Structure of [Cu(CH₃NO₂)₅(NO)][PF₆]₂.

bond in this complex is found to be much weaker as compared to other Cu—NO complexes. The application of a vacuum at room temperature results into the release of nitric oxide gas from the complex. This complex has a promising potential application as a nitric oxide donor.

6.4 Dinitrosylmolybdenum(0) with Schiff Bases

Hexacoordinated mixed-ligand dinitrosyl complexes of molybdenum(0) of the composition [Mo(NO)$_2$(L)(OH)] with a variety of Schiff base ligands have been reported [3]. Such complexes are readily obtained by a single-step reaction between molybdate(VI) and the ligand in presence of hydroxylaminehydrochloride in a dimethylformamide solvent. A *cis*-octahedral structure has been observed in these complexes, as shown in Figure 12.

The identification of the coordinated nitrosyl group in the complexes can be completed by decomposition of the complex with KOH, followed by acidification with acetic acid. The filtrate, thus obtained when treated with a few drops of the Griess Reagent (sulfanilamide and 1-napthylethylenediamine dihydrochloride in phosphoric acid), shows a pink colour, indicating the coordination of nitrite ion in the complex. The pink colour is observed due to

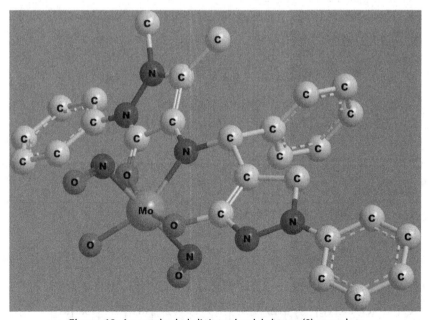

Figure 12 An octahedral dinitrosylmolybdenum(0) complex.

the formation of an azo dye between the reagents of Griess reagents. The amine group of sulfanilamide is diazotized by the nitrite ion, which facilitates the dye formation. The ν N$-$O in the infrared spectra of these compounds are observed in the range of 1770 and 1650 cm^{-1}. Both the nitrosyl ligands exhibit a bond angle M$-$N$-$O in the range of 120–140 °C.

7. EFFECTIVE ATOMIC NUMBER (EAN) RULE

The EAN for metallic nitrosyls can be calculated in a manner similar to that of carbonyls. In metallic nitrosyls, the metals exhibit a variable oxidation state. Hence, instead of considering the nuclear charge Z for the calculation, $Z + n$ should be used. Here, 'n' indicates the number of electrons gained or lost by the metal to achieve the oxidation state observed in the nitrosyl complex.

The EAN for $[Co^-(CO)_3(NO^+)]^0$ can be calculated as follows:

$$EAN = Z + n + a + b + c$$

Here,

Co ($Z = 27$) is in -1 oxidation state, which is achieved by gaining one electron. Hence, $n = +1$. Moreover, all the four ligand groups (three carbonyls and one nitrosyl) are terminal groups, with each donating an electron pair, i.e. $a = 4 \times 2 = 8$.

There are no bridge bonds or M$-$M bonds in the complex. Therefore, both b and c equals zero.

Thus,

$$EAN = 27 + 1 + 8 + 0 + 0 = 36 \text{ [Kr]}$$

Thus, $[Co^-(CO)_3(NO^+)]^0$ obeys the EAN rule. Calculation of EAN for selected nitrosyl complexes are listed in Table 3.

Table 3 Calculation of effective atomic number (EAN) in some nitrosyls

Metal nitrosyl	$Z + n$	a	b	c	EAN $= Z + n +$ $a + b + c$
$[Co^-(CO)_3(NO^+)]^0$	$27 + 1$	$4 \times 2 = 8$	0	0	36 [Kr]
$[Fe^{2-}(CO)_2(NO^+)_2]^0$	$26 + 2$	$4 \times 2 = 8$	0	0	36 [Kr]
$[Mn^{3-}(CO)(NO^+)_3]^0$	$25 + 3$	$4 \times 2 = 8$	0	0	36 [Kr]
$[Cr^+(NO^+)(CN)_5]^{-3}$	$24 - 1$	$6 \times 2 = 12$	0	0	35
$[Fe^+(NO^+)(H_2O)_5]^{-2}$	$26 - 1$	$6 \times 2 = 12$	0	0	37
$[Co^{+3}(NO^-)(CN)_5]^{-3}$	$27 - 3$	$6 \times 2 = 12$	0	0	36 [Kr]
$[Fe^{-2}(NO^+)_3(NO^-)]^{-2}$	$26 + 2$	$4 \times 2 = 8$	0	0	36 [Kr]

8. APPLICATIONS OF SOME METALLIC NITROSYLS

8.1 As a Catalyst for Ring-Opening Metathesis Polymerization (ROMP) Reaction

Cationic rhenium dinitrosyl bisphosphine complexes reportedly catalyse the ring-opening metathesis polymerization (ROMP) of highly strained nonfunctionalized cyclic olefins [4], as shown in Figure 13.

8.2 As Wound Healers Dinitrosyl Iron Complexes

Composites of a collagen matrix and dinitrosyl iron complexes with glutathione, when applied in a rat skin wound, induced the acceleration of wound closure in 21 days instead of 23 days [5]. It is suggested that the positive effect of the material on wound healing is based on the release of NO.

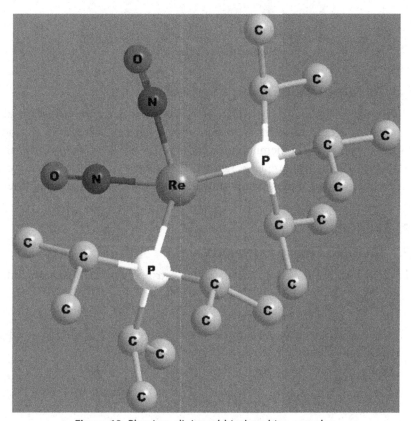

Figure 13 Rhenium dinitrosyl bisphosphine complex.

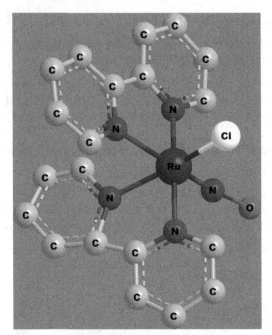

Figure 14 *Cis*-[Ru(bpy)(NO)Cl].

8.3 As Nitric Oxide Delivery Systems

Polypyridine ruthenium complexes are widely studied as nitric oxide delivery systems [6]. They are useful as artificial light-harvesting systems for technological purposes and as chemiluminescence probes. *Cis*-[Ru(bpy)(NO)Cl] and [Ru(tby)(NO)(bpy)] complexes, shown in Figures 14 and 15, constitute a representative class of these compounds capable of functioning as drug carriers.

9. EXERCISES

9.1 Multiple Choice Questions

1. Which of the following species is isoelectronic with carbon monoxide?
   (a) NO^+ (b) NO
   (c) NO^- (d) none of these

2. How many electrons does nitric oxide donate when bonded to metal as NO^+?
   (a) 1   (b) 2
   (c) 3   (d) 4

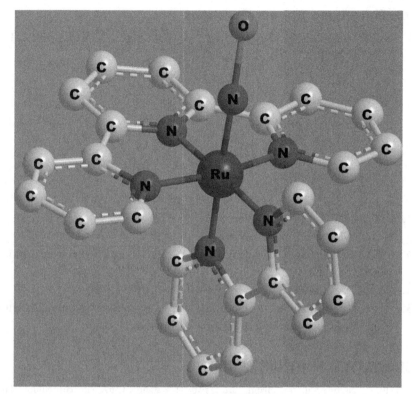

Figure 15 [Ru(tby)(NO)(bpy)].

3. The expected value for $\angle M{-}N{-}O$ in complexes containing NO^+ ligand is _____.
   (a) 180   (b) 120
   (c) 90   (d) uncertain

4. The expected value for $\angle M{-}N{-}O$ in complexes containing NO^- ligand is _____.
   (a) 180   (b) 120
   (c) 90   (d) uncertain

5. In which test is nitroso ferrous sulphate produced?
   (a) brown ring test   (b) Simon's test
   (c) azo dye test   (d) Fehling's test

6. Sodium nitroprusside is useful in the treatment of _____.
   (a) hepatitis   (b) hypertension
   (c) cancer   (d) diabetes

9.2 Short/Long Answer Questions

1. Using the donor pair method, perform the electron count for $[Cr(CN)_5(NO)]^{4-}$, considering
   (i) Cr—N—O angle $= 120°$ (ii) Cr—N—O angle $= 180°$.

Solution:

Bent (120°):			Linear (180°):		
	$5CN^-$	$10e^-$		$5CN^-$	$10e^-$
	$1NO^-$	$2e^-$		$1NO^+$	$2e^-$
	Cr^{2+}	$4e^-$		Cr^0	$6e^-$
	Total	$16e^-$		Total	$18e^-$

2. For an 18-electron complex ion, $[Fe(CN)_5(NO)]^{2-}$, what is the expected M—N—O angle? Why?
3. 'NO^+ is a three-electron donor while NO^- is a one-electron donor': Justify.
4. Discuss the infrared spectra of $[CoCl_2(NO)(PCH_3Ph_2)_2]$.
5. Why is it difficult to correlate the N—O stretching frequencies with $\angle$M—N—O in the case of nitrosyl complexes? Explain.

SUGGESTED FURTHER READINGS

The topics discussed in this chapter are a part of a standard graduate curriculum. The majority of the textbooks with titles related to coordination chemistry and organometallic chemistry can act as a source of further reading. Moreover, there are several web resources useful for further learning. Some of them are listed below:

https://en.wikipedia.org/wiki/Metal_nitrosyl_complex
http://www.britannica.com/science/metal-nitrosyl
http://www.chem.ucla.edu/~bacher/CHEM174/hints/Iron_Nitrosyl.html
http://nptel.ac.in/courses/104106064/

REFERENCES

The following books and research articles contain advanced topics and nuances in the chemistry of metal nitrosyl complexes:
[1] Enemark JH, Feltham RD. Principles of structure, bonding, and reactivity for metal nitrosyl complexes. Coord Chem Rev 1974;13(4):339–406.
[2] Wright AM, Wu G, Hayton TW. Structural characterization of a copper nitrosyl complex with a {CuNO}10 configuration. J Am Chem Soc 2010;132(41): 14336–7.
[3] Maurya RC, et al. Metal nitrosyl complexes of bioinorganic, catalytic, and environmental relevance: a novel single-step synthesis of dinitrosylmolybdenum(0) complexes of {Mo(NO)2}6 electron configuration involving Schiff bases derived from 4-acyl-3-methyl-1-phenyl-2-pyrazolin-5-one and 4-aminoantipyrine, directly from molybdate(VI) and their characterization. J Mol Struct 2006;798(1–3):89–101.

[4] Frech CM, et al. Unprecedented ROMP activity of low-valent rhenium–nitrosyl complexes: mechanistic evaluation of an electrophilic olefin metathesis system. Chemistry 2006;12(12):3325–38.

[5] Shekhter AB, et al. Dinitrosyl iron complexes with glutathione incorporated into a collagen matrix as a base for the design of drugs accelerating skin wound healing. Eur J Pharm Sci 2015;78:8–18.

[6] deBoer TR, Mascharak PK. Chapter three-recent progress in photoinduced no delivery with designed ruthenium nitrosyl complexes. In: Rudi van E, José AO, editors. Advances in inorganic chemistry. Academic Press; 2015. p. 145–70.

INDEX

Note: Page numbers followed by "f" and "t" indicate figures and tables, respectively.